Succeed

Eureka Math®
Grade 2
Modules 4 & 5

Published by Great Minds®.

Copyright © 2018 Great Minds®.

Printed in the U.S.A.
This book may be purchased from the publisher at eureka-math.org.
BAB 10 9 8 7 6 5 4 3

ISBN 978-1-64054-085-9

G2-M4-M5-S-06.2018

Learn ◆ Practice ◆ Succeed

Eureka Math® student materials for *A Story of Units®* (K–5) are available in the *Learn, Practice, Succeed* trio. This series supports differentiation and remediation while keeping student materials organized and accessible. Educators will find that the *Learn, Practice,* and *Succeed* series also offers coherent—and therefore, more effective—resources for Response to Intervention (RTI), extra practice, and summer learning.

Learn

Eureka Math Learn serves as a student's in-class companion where they show their thinking, share what they know, and watch their knowledge build every day. *Learn* assembles the daily classwork—Application Problems, Exit Tickets, Problem Sets, templates—in an easily stored and navigated volume.

Practice

Each *Eureka Math* lesson begins with a series of energetic, joyous fluency activities, including those found in *Eureka Math Practice.* Students who are fluent in their math facts can master more material more deeply. With *Practice,* students build competence in newly acquired skills and reinforce previous learning in preparation for the next lesson.

Together, *Learn* and *Practice* provide all the print materials students will use for their core math instruction.

Succeed

Eureka Math Succeed enables students to work individually toward mastery. These additional problem sets align lesson by lesson with classroom instruction, making them ideal for use as homework or extra practice. Each problem set is accompanied by a Homework Helper, a set of worked examples that illustrate how to solve similar problems.

Teachers and tutors can use *Succeed* books from prior grade levels as curriculum-consistent tools for filling gaps in foundational knowledge. Students will thrive and progress more quickly as familiar models facilitate connections to their current grade-level content.

Students, families, and educators:

Thank you for being part of the *Eureka Math*® community, where we celebrate the joy, wonder, and thrill of mathematics.

Nothing beats the satisfaction of success—the more competent students become, the greater their motivation and engagement. The *Eureka Math Succeed* book provides the guidance and extra practice students need to shore up foundational knowledge and build mastery with new material.

What is in the Succeed *book?*

Eureka Math Succeed books deliver supported practice sets that parallel the lessons of *A Story of Units*®. Each *Succeed* lesson begins with a set of worked examples, called *Homework Helpers*, that illustrate the modeling and reasoning the curriculum uses to build understanding. Next, students receive scaffolded practice through a series of problems carefully sequenced to begin from a place of confidence and add incremental complexity.

How should Succeed *be used?*

The collection of *Succeed* books can be used as differentiated instruction, practice, homework, or intervention. When coupled with *Affirm*®, *Eureka Math*'s digital assessment system, *Succeed* lessons enable educators to give targeted practice and to assess student progress. *Succeed*'s perfect alignment with the mathematical models and language used across *A Story of Units* ensures that students feel the connections and relevance to their daily instruction, whether they are working on foundational skills or getting extra practice on the current topic.

Where can I learn more about Eureka Math *resources?*

The Great Minds® team is committed to supporting students, families, and educators with an ever-growing library of resources, available at eureka-math.org. The website also offers inspiring stories of success in the *Eureka Math* community. Share your insights and accomplishments with fellow users by becoming a *Eureka Math* Champion.

Best wishes for a year filled with Eureka moments!

Jill Diniz

Jill Diniz
Director of Mathematics
Great Minds

Contents

Module 4: Addition and Subtraction Within 200 with Word Problems to 100

Module 5: Addition and Subtraction Within 1,000 with Word Problems to 100

Topic C: Strategies for Decomposing Tens and Hundreds Within 1,000

Topic D: Student Explanations for Choice of Solution Methods

Grade 2
Module 4

1. Complete each more or less statement.

    a. 1 less than 46 is **45.**

    b. 48 is 10 more than **38.**

    c. **63** is 10 less than 73.

    d. 39 is 1 *less than* 40.

    > I can use place value language to explain the change. 1 more and 10 more are the same as adding. 1 less and 10 less are the same as subtracting.

2. Complete each pattern, and write the rule.

    a. 33, 34, **35, 36,** 37   Rule: **1** *more*

    b. 43, **33,** 23, **13,** 3   Rule: **10** *less*

    c. **43,** 42, 41, 40, **39**   Rule: **1** *less*

    > I study the numbers and look for the more or less pattern. I know 34 is 1 more than 33, so the rule is 1 more.

    > 40 is 1 less than 41, so the rule is 1 less.

3. Label each statement as true or false.

    a. 1 more than 43 is the same as 1 less than 45.   *True.*

    b. 10 less than 28 is the same as 1 more than 16.   *False.*

    > I know 1 more than 43 is 44, and 1 less than 45 is 44, so this statement is true.
    >
    > 10 less than 28 is 18, and 1 more than 16 is 17, so this statement is false.

4. Below is a chart of fruit in Gloria's basket.

    Use the following to complete the chart.

    • Gloria has 1 more banana than the number of apples.

    • Gloria has 10 fewer oranges than the number of pears.

    > I can use what I know about number patterns to complete the chart. 1 more than 19 is 20, so there are 20 bananas. 10 fewer than 21 is 11, so there are 11 oranges.

Fruit	Number of Fruit
Apples	19
Pears	21
Bananas	**20**
Oranges	**11**

Lesson 1:   Relate 1 more, 1 less, 10 more, and 10 less to addition and subtraction of 1 and 10.

© 2018 Great Minds®. eureka-math.org

3

Name _____   Date _____

1. Complete each *more* or *less* statement.

    a.   1 more than 37 is _____.   b.   10 more than 37 is _____.

    c.   1 less than 37 is _____.   d.   10 less than 37 is _____.

    e.   58 is 10 more than _____.   f.   29 is 1 less than _____.

    g.   _____ is 10 less than 45.   h.   _____ is 1 more than 38.

    i.   49 is _____ than 50.   j.   32 is _____ than 22.

2. Complete each pattern and write the rule.

    a.   44, 45, _____, _____, 48   Rule: _____

    b.   44, _____, 24, _____, 4   Rule: _____

    c.   44, _____, _____, 74, 84   Rule: _____

    d.   _____, 43, 42, _____, 40   Rule: _____

    e.   _____, _____, 44, 34, _____   Rule: _____

    f.   41, _____, _____, 38, 37   Rule: _____

3. Label each statement as true or false.

   a. 1 more than 36 is the same as 1 less than 38. _____

   b. 10 less than 47 is the same as 1 more than 35. _____

   c. 10 less than 89 is the same as 1 less than 90. _____

   d. 10 more than 41 is the same as 1 less than 43. _____

4. Below is a chart of balloons at the county fair.

Color of Balloons	Number of Balloons
Red	59
Yellow	61
Green	65
Blue	
Pink	

   a. Use the following to complete the chart and answer the question.

      ▪ The fair has 1 more blue than red balloons.

      ▪ There are 10 fewer pink than yellow balloons.

      Are there more blue or pink balloons?

   b. If 1 red balloon pops and 10 red balloons fly away, how many red balloons are left? Use the arrow way to show your work.

EUREKA
MATH

1. Solve using place value strategies. Use the arrow way, number bonds, or mental math, and record your answers.

   a. $48 + 30 = \textbf{78}$

       /\
       40 8 $40 + 30 = 70$
                          $70 + 8 = 78$

   > To solve 48 + 30, I think 30 more than 48. I just add like units! 30 + 40 = 70, and 70 + 8 = 78. I can draw a number bond to show my thinking.

   b. $27 + 20 = 47$

       $27 \xrightarrow{+20} 47$

   > To solve 27 + __ = 47, I count by tens from 27 to 47. I can use the arrow way to show my thinking.

2. Find each sum. Then use >, <, or = to compare.

   a. $43 + 20 < 30 + 53$

   b. $29 + 50 > 40 + 19$

   > 20 more than 43 is 63, and 30 more than 53 is 83, so 63 is less than 83.

3. Solve using place value strategies.

   a. $35 - 20 = \textbf{15}$

   b. $46 - \textbf{20} = 26$

   > I can draw or solve in my head using place value thinking. 3 tens 5 ones − 2 tens is 1 ten 5 ones, so 35 − 20 = 15.

4. Complete each more than or less than statement.

   a. 30 less than 78 is **48**.   b. 45 more than 30 is **75**.

   c. 20 less than **68** is 48.   d. 40 more than **22** is 62.

   > 20 less than what number is 48? I can count on to solve! 48, 58, 68. 20 less than 68 is 48.

   > To solve, I just add like units! 45 more than 30 is the same as 45 + 30. 40 + 30 = 70, and 70 + 5 = 75.

5. There were 53 papers in the bin after math class. There were 20 papers in the bin before math class. How many papers were added during math class? Use the arrow way to show your simplifying strategy.

   $$20 \xrightarrow{+10} 30 \xrightarrow{+10} 40 \xrightarrow{+10} 50 \xrightarrow{+3} 53$$

   33 papers were added to the bin during math class.

   > I can start at 20 and count on by tens to 50, and then just add 3 ones to get to 53.

Name _____   Date _____

1. Solve using place value strategies. Use scrap paper to show the arrow way or number bonds, or just use mental math, and record your answers.

a. 2 tens + 3 tens = _____ tens 20 + 30 = _____ 2 tens 4 ones + 3 tens= ___ tens ___ ones 24 + 30 = _____	b. 5 tens + 4 tens = _____ tens 50 + 40 = _____ 5 tens 9 ones + 4 tens = ___ tens ___ ones 59 + 40 = _____

c. 28 + 40 = _____   18 + 30 = _____   60 + 38 = _____

d. 30 + 25 = _____   35 + 50 = _____   15 + 20 = _____

e. 37 + _____ = 47   _____ + 27 = 57   17 + _____ = 87

f. _____ + 22 = 62   29 + _____ = 79   11 + _____ = 91

2. Find each sum. Then use >, <, or = to compare.

a. 23 + 40 _____ 20 + 33   d. 64 + 10 _____ 49 + 20

b. 50 + 18 _____ 48 + 20   e. 70 + 21 _____ 18 + 80

c. 19 + 60 _____ 39 + 30   f. 35 + 50 _____ 26 + 60

3. Solve using place value strategies.

| a.   6 tens – 2 tens = _____ tens

     60 – 20 = _____

6 tens 3ones – 3 tens = ___ tens ___ones

     63 – 30 = _____ | b.   8 tens – 5 tens = _____ tens

     80 – 50 = _____

8 tens 9 ones – 5 tens = ___ tens ___ones

     89 – 50 = ___ |

c.   55 – 20 = _____   75 – 30 = _____   85 – 50 = _____

d.   72 – _____ = 22   49 – _____ = 19   88 – _____ = 28

e.   67 – _____ = 47   71 – _____ = 51   99 – _____ = 69

4. Complete each more than or less than statement.

a.   20 less than 58 is _____.

b.   36 more than 40 is _____.

c.   40 less than _____ is 28.

d.   50 more than _____ is 64.

5. There were 68 plates in the sink at the end of the day. There were 40 plates in the sink at the beginning of the day. How many plates were added throughout the day? Use the arrow way to show your simplifying strategy.

EUREKA MATH

1. Solve using the arrow way.

   a. $48 + 30 = 78$

   $$48 \xrightarrow{+30} 78$$

   I can use the arrow way to show my thinking. 30 more than 48 is 78. I just add like units, 4 tens plus 3 tens is 7 tens. The 8 ones stay the same.

   $48 + 31 = 79$

   $$48 \xrightarrow{+30} 78 \xrightarrow{+1} 79$$

   To add 31, I add 3 tens, and then just add 1 more.

   $48 + 29 = 77$

   $$48 \xrightarrow{+30} 78 \xrightarrow{-1} 77$$

   Adding 29 is adding 1 less than 30. I add 3 tens, and then just take 1 away.

   b. $57 - 40 = 17$

   $$57 \xrightarrow{-40} 17$$

   40 less than 57 is 17. I just subtract like units. 5 tens minus 4 tens is 1 ten. The 7 ones stay the same.

   $57 - 41 = 16$

   $$57 \xrightarrow{-40} 17 \xrightarrow{-1} 16$$

   To subtract 41, I subtract 4 tens and then just subtract 1 one.

   $57 - 39 = 18$

   $$57 \xrightarrow{-40} 17 \xrightarrow{+1} 18$$

   Subtracting 39 is subtracting 1 less than 40. I subtract 4 tens and then just add 1 one.

   The first problem, $57 - 40$, helps me solve the last problem, $57 - 39$. Subtracting 40 is easy, but that's 1 more than I'm supposed to take away, so I have to add 1 back, which means the answer is 18.

EUREKA MATH

Lesson 3:   Add and subtract multiples of 10 and some ones within 100.

11

2. Solve using the arrow way, number bonds, or mental math.

$$43 + 20 = \mathbf{63}$$

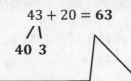

> I can solve mentally, in my head! 20 more than 43 is 63. A number bond is another way I can show how I add like units.

EUREKA
MATH®

Name _____   Date_____

1. Solve using the arrow way. The first set is done for you.

<table>
<tr><td>

a.

$67 + 20 =$ __87__

$67 \xrightarrow{+20}$ __87__

$67 + 21 =$ __88__

$67 \xrightarrow{+20}$ __87__ $\xrightarrow{+1}$ __88__

$67 + 19 =$ __86__

$67 \xrightarrow{+20}$ __87__ $\xrightarrow{-1}$ __86__

</td><td>

b.

$56 + 40 =$ ____

$56 + 41 =$ ____

$56 + 39 =$ ____

</td></tr>
<tr><td>

c.

$68 - 40 =$ ____

$68 - 41 =$ ____

$68 - 39 =$ ____

</td><td>

d.

$87 - 50 =$ ____

$87 - 51 =$ ____

$87 - 49 =$ ____

</td></tr>
</table>

EUREKA MATH

Lesson 3:   Add and subtract multiples of 10 and some ones within 100.

13

© 2018 Great Minds®. eureka-math.org

2. Solve using the arrow way, number bonds, or mental math. Use scrap paper if needed.

a.	b.	c.
48 - 20 = _____	86 - 50 = _____	37 + 40 = _____
48 - 21 = _____	86 - 51 = _____	37 + 41 = _____
48 - 19 = _____	86 - 49 = _____	37 + 39 = _____
d.	e.	f.
62 + 30 = _____	77 - 40 = _____	28 + 50 = _____
62 + 31 = _____	77 - 41 = _____	28 + 51 = _____
62 + 29 = _____	77 - 39 = _____	28 + 49 = _____

3. Marcy had $84 in the bank. She took $39 out of her account. How much does she have in her account now?

4. Brian has 92 cm of rope. He cuts off a piece 49 cm long to tie a package.
   a. How much rope does Brian have left?

   b. To tie a different package, Brian needs another piece of rope that is 8 cm shorter than the piece he just cut. Does he have enough rope left?

Lesson 3:   Add and subtract multiples of 10 and some ones within 100.

EUREKA
MATH

1. Solve. Draw and label a tape diagram to subtract 10, 20, 30, 40, etc.

$23 - 9 = 24 - 10 = 14$

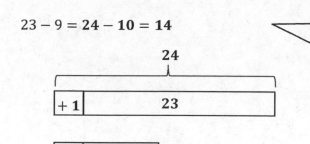

It is easier to subtract a multiple of 10. 9 is very close to 10; it just needs 1 more. I can add 1 to both numbers to make it easier to subtract, and the difference will not change. A tape diagram helps me show my strategy.

2. Solve. Draw a number bond to add 10, 20, 30, 40, etc.

$38 + 53 = 40 + 51 = 91$
$\quad\quad /\backslash$
$\quad\quad 2\ 51$

It is easier to add a multiple of 10. 38 is very close to a 10, it just needs 2 more. I can break apart 53 into 2 and 51 to get the 2 out. 38 plus 2 is 40. Now I just add what is left; 40 plus 51 is 91.

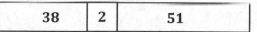

I can also show this with a tape diagram! This helps me see that if I take 2 from 53 and give it to 38, I get 40 + 51.

Name _____   Date _____

1. Solve. Draw and label a tape diagram to subtract 10, 20, 30, 40, etc.

   a. 17 - 9 = ___18 - 10___ = _____

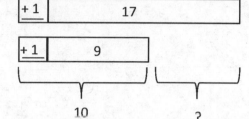

   b. 33 - 19 = _____ = _____

   c. 60 - 29 = _____ = _____

   d. 56 - 38 = _____ = _____

EUREKA MATH

Lesson 4:   Add and subtract multiples of 10 and some ones within 100.

17

© 2018 Great Minds®. eureka-math.org

2. Solve. Draw a number bond to add 10, 20, 30, 40, etc.

   a.   28 + 43 = ___30 + 41___ = _____

            /\
           2 41

   b. 49 + 26 = _____ = _____

   c. 43 + 19 = _____ = _____

   d. 67 + 28 = _____ = _____

3. Kylie has 28 more oranges than Cynthia. Kylie has 63 oranges. How many oranges does Cynthia have? Draw a tape diagram or number bond to solve.

EUREKA
MATH

Solve and show your strategy.

1. There are 38 fewer green apples in the orchard than red apples. There are 62 green apples in the orchard. How many red apples are there?

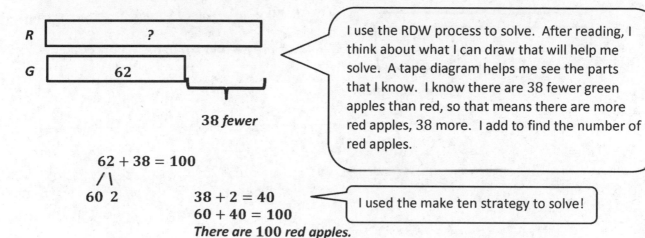

I use the RDW process to solve. After reading, I think about what I can draw that will help me solve. A tape diagram helps me see the parts that I know. I know there are 38 fewer green apples than red, so that means there are more red apples, 38 more. I add to find the number of red apples.

$$62 + 38 = 100$$
/\
60 2

$$38 + 2 = 40$$
$$60 + 40 = 100$$
There are 100 red apples.

I used the make ten strategy to solve!

2. Oscar has two baskets of toys. The red basket has 27 toys. The yellow basket has 29 more toys than the red basket.

   a. How many toys are in the yellow basket?

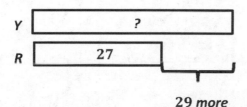

29 more

The yellow basket has 29 more than the red basket. I add to find 29 more than 27. I can use the make ten strategy here, too!

$$27 + 29 = 56$$
/\
26 1

$$29 + 1 = 30$$
$$26 + 30 = 56$$
The yellow basket has 56 toys.

b. Oscar gave 18 toys from the yellow basket to his younger brother. How many toys are left in the yellow basket?

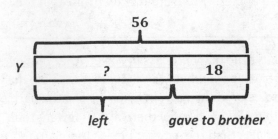

$56 - 18 = 38$

$/\ $

36 20 $20 - 18 = 2$

              $36 + 2 = 38$

There are 38 toys left in the yellow basket.

Lesson 5:   Solve one- and two-step word problems within 100 using strategies based on place value.

EUREKA MATH

Name _____ Date _____

Solve and show your strategy.

1. 38 markers were in the bin. Chase added the 43 markers that were on the floor to the bin. How many markers are in the bin now?

2. There are 29 fewer big stickers on the sticker sheet than little stickers. There are 62 little stickers on the sheet. How many big stickers are there?

Lesson 5:   Solve one- and two-step word problems within 100 using strategies based on place value.

© 2018 Great Minds®. eureka-math.org

21

3. Rose has 34 photos in a photo album and 41 photos in a box. How many photos does Rose have?

4. Halle has two ribbons. The blue ribbon is 58 cm. The green ribbon is 38 cm longer than the blue ribbon.

a. How long is the green ribbon?

b. Halle uses 67 cm of green ribbon to wrap a present. How much green ribbon is left?

Lesson 5:   Solve one- and two-step word problems within 100 using strategies based on place value.

EUREKA MATH

5. Chad bought a shirt for $19 and a pair of shoes for $28 more than the shirt.

   a. How much was the pair of shoes?

   b. How much money did Chad spend on the shirt and shoes?

   c. If Chad had $13 left over, how much money did Chad have before buying the shirt and shoes?

Lesson 5:   Solve one- and two-step word problems within 100 using strategies
based on place value.

© 2018 Great Minds®. eureka-math.org

23

1. Solve the following problems using your place value chart and place value disks. Compose a ten, if needed. Think about which ones you can solve mentally, too!

$34 + 25 =$ ___59___

> I can solve this one mentally! I just add like units. 3 tens and 2 tens is 5 tens. 4 ones and 5 ones is 9 ones. Altogether that makes 5 tens 9 ones, or 59.

$34 + 28 =$ ___62___

> I can use my chart and place value disks to solve this problem.

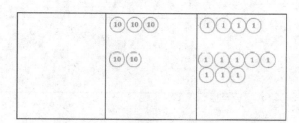

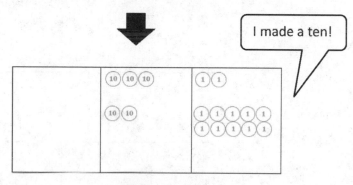

> I made a ten!

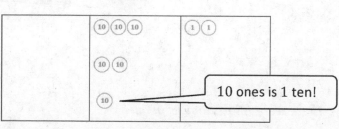

> So, $34 + 28 = 62$.

> 10 ones is 1 ten!

EUREKA MATH

Lesson 6:   Use manipulatives to represent the composition of 10 ones as 1 ten with two-digit addends.

© 2018 Great Minds®. eureka-math.org

25

2. Solve using a place value chart.

Marty used 28 toothpicks for his art project, and 37 were left in the box. How many toothpicks were there in all?

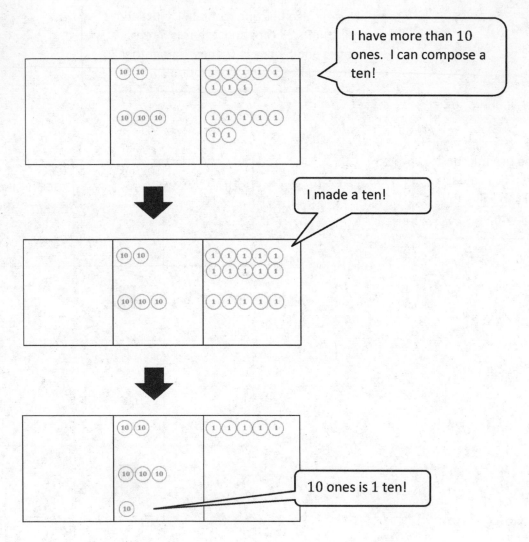

$28 + 37 = 65$

There were 65 toothpicks in all.

Lesson 6:   Use manipulatives to represent the composition of 10 ones as 1 ten with two-digit addends.

© 2018 Great Minds®. eureka-math.org

EUREKA MATH

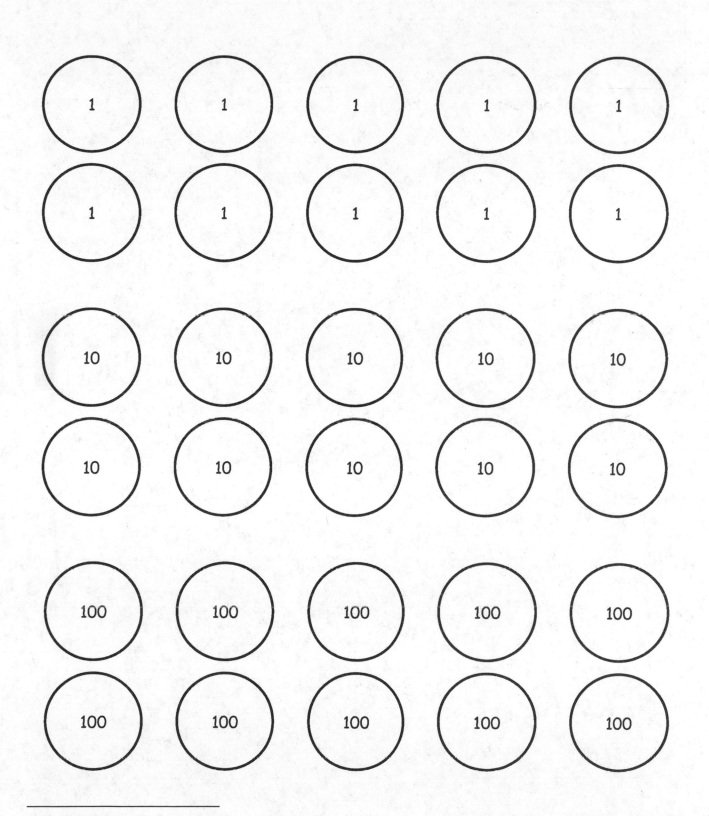

place value disks

Lesson 6:   Use manipulatives to represent the composition of 10 ones as 1 ten
with two-digit addends.

29

© 2018 Great Minds®. eureka-math.org

unlabeled hundreds place value chart - from Lesson 18

Lesson 6:   Use manipulatives to represent the composition of 10 ones as 1 ten
             with two-digit addends.

© 2018 Great Minds®. eureka-math.org

31

1. Solve the following problems using the vertical form, your place value chart, and place value disks. Bundle a ten, if needed. Think about which ones you can solve mentally, too!

a. $33 + 7 = 40$

> I can solve this one mentally! I know 3 ones plus 7 ones is 1 ten, and 30 plus 10 is 40.

b. $36 + 57 = 93$

> I can use my chart and place value disks to solve.

> I can write it in vertical form as I model it with my place value disks.

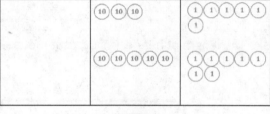

$$\begin{array}{r} 3\ \ 6 \\ +\ 5\ \ 7 \\ \hline \end{array}$$

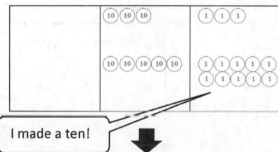

$$\begin{array}{r} 3\ \ 6 \\ +\ 5\ \ 7 \\ \underline{\ \ 1\quad} \\ 3 \end{array}$$

> I have 13 ones, or 1 ten 3 ones. I show the ten, using new groups below, on the line below the tens place.

I made a ten!

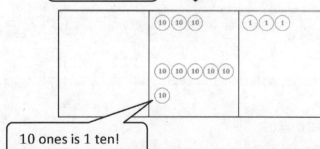

$$\begin{array}{r} 3\ \ 6 \\ +\ 5\ \ 7 \\ \underline{\ \ 1\quad} \\ 9\ \ 3 \end{array}$$

> Now I just add the tens! 3 tens plus 5 tens is 8 tens, and 1 more ten is 9 tens. So 36 plus 57 is 93.

10 ones is 1 ten!

EUREKA MATH

Lesson 7:   Relate addition using manipulatives to a written vertical method.

33

© 2018 Great Minds®. eureka-math.org

2. Add the bottom numbers to find the missing number above it.

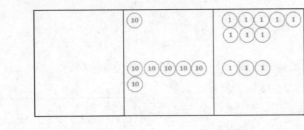

$$\begin{array}{r} 1\ 9 \\ +\ 3\ 5 \\ \hline {\scriptstyle 1} \\ 5\ 4 \end{array}$$

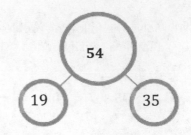

I can solve using my place value disks and vertical form or the make ten strategy!

$19 + 35 = 54$
/\
1 34

3. Jen's ribbon is 18 centimeters longer than her desk. The length of her desk is 63 centimeters.

a. What is the length of Jen's ribbon?

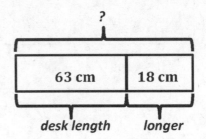

63 cm   18 cm

desk length   longer

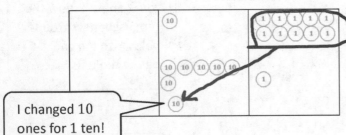

$$\begin{array}{r} 1\ 8 \\ +\ 6\ 3 \\ \hline {\scriptstyle 1} \\ 8\ 1 \end{array}$$

Jen's ribbon is 81 centimeters.

I changed 10 ones for 1 ten!

b. The length of Jen's desk is 20 centimeters shorter than the length of her teacher's desk. How long is her teacher's desk?

$63 + 20 = 83$

The teacher's desk is 83 centimeters long.

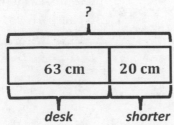

?

63 cm   20 cm

desk   shorter

Lesson 7:   Relate addition using manipulatives to a written vertical method.

EUREKA MATH

Solve vertically. Draw and bundle place value disks on the place value chart.

1. $27 + 45 = \mathbf{72}$

$$
\begin{array}{r}
2\ \ 7 \\
+\ 4\ \ 5 \\
\scriptstyle 1\ \ \ \\
\hline
7\ \ 2
\end{array}
$$

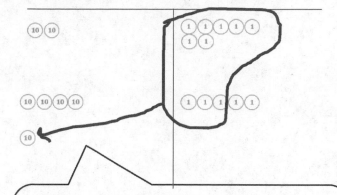

I show each step I make with the place value disks vertically using new groups below.

I draw place value disks to show each addend. 7 ones plus 5 ones is 12 ones, or 1 ten 2 ones. I bundle 10 ones to make 1 ten. Now I just add the tens. 2 tens plus 4 tens plus 1 more ten is 7 tens. So 27 plus 45 is 72.

2.   Santiago counted the number of people on two buses. Bus 1 had 29 people, and bus 2 had 34 people. How many people were on the two buses?

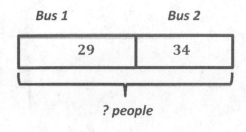

Bus 1	Bus 2
29	34

? people

$$
\begin{array}{r}
2\ \ 9 \\
+\ 3\ \ 4 \\
\scriptstyle 1\ \ \ \\
\hline
6\ \ 3
\end{array}
$$

63 people were on the two buses.

1. Solve using the algorithm. Draw and bundle chips on the place value chart.

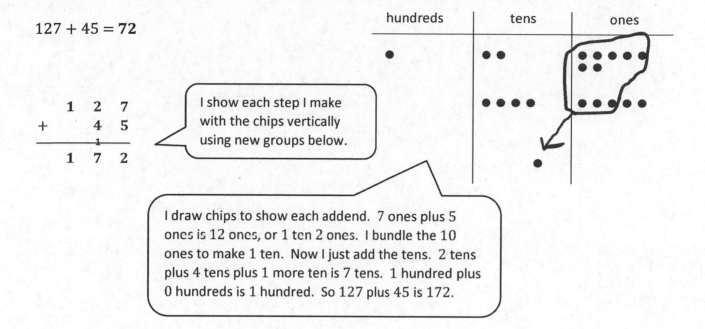

$127 + 45 = \mathbf{72}$

```
    1  2  7
 +     4  5
    ‾‾1‾‾‾‾‾
    1  7  2
```

I show each step I make with the chips vertically using new groups below.

I draw chips to show each addend. 7 ones plus 5 ones is 12 ones, or 1 ten 2 ones. I bundle the 10 ones to make 1 ten. Now I just add the tens. 2 tens plus 4 tens plus 1 more ten is 7 tens. 1 hundred plus 0 hundreds is 1 hundred. So 127 plus 45 is 172.

2. Solve using the algorithm. Write a number sentence for the problem modeled on the place value chart.

I can count to find the first addend: 100, 110, 120, 130, 140, 141, 142, 143, 144, 145. The first addend is 145. Now I count to find the second addend: 10, 20, 21, 22, 23, 24, 25, 26, 27, 28. The second addend is 28.

hundreds	tens	ones

```
    1  4  5
 +     2  8
    ‾‾1‾‾‾‾‾
    1  7  3
```

EUREKA MATH

Lesson 9:   Use math drawings to represent the composition when adding a
             two-digit to a three-digit addend.

© 2018 Great Minds®. eureka-math.org

41

Name _____   Date _____

1. Solve using the algorithm. Draw and bundle chips on the place value chart.

    a.   127 + 14 = _____

hundreds	tens	ones

    b.   135 + 46 = _____

hundreds	tens	ones

    c.   108 + 37 = _____

hundreds	tens	ones

EUREKA MATH

Lesson 9:   Use math drawings to represent the composition when adding a two-digit to a three-digit addend.

© 2018 Great Minds®. eureka-math.org

43

2. Solve using the algorithm. Write a number sentence for the problem modeled on the place value chart.

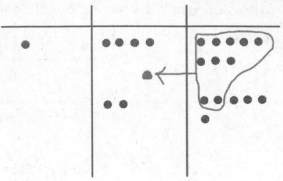

3. Jane made 48 lemon bars and 23 cookies.

    a. How many lemon bars and cookies did Jane make?

hundreds	tens	ones

    b. Jane made 19 more lemon bars. How many lemon bars does she have?

hundreds	tens	ones

Lesson 9:   Use math drawings to represent the composition when adding a
            two-digit to a three-digit addend.

© 2018 Great Minds®. eureka-math.org

EUREKA
MATH

1. Solve using the algorithm. Draw and bundle chips on the place value chart.

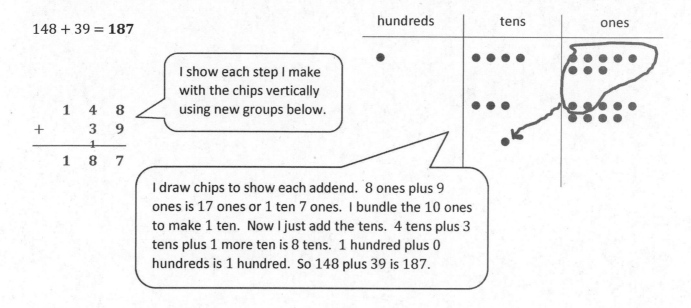

$148 + 39 = \mathbf{187}$

$$\begin{array}{r} 1\ 4\ 8 \\ +\quad 3\ 9 \\ \underline{1} \\ 1\ 8\ 7 \end{array}$$

I show each step I make with the chips vertically using new groups below.

I draw chips to show each addend. 8 ones plus 9 ones is 17 ones or 1 ten 7 ones. I bundle the 10 ones to make 1 ten. Now I just add the tens. 4 tens plus 3 tens plus 1 more ten is 8 tens. 1 hundred plus 0 hundreds is 1 hundred. So 148 plus 39 is 187.

2. Frankie spilled ink on his paper. Can you figure out what problem he was given by looking at his work?

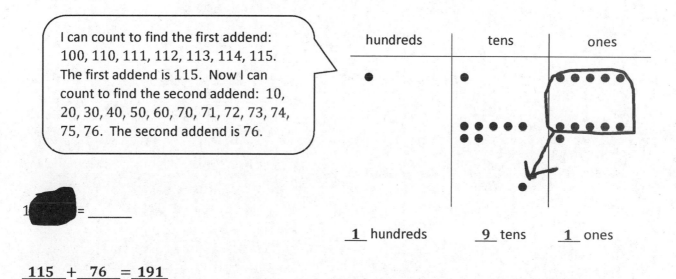

I can count to find the first addend: 100, 110, 111, 112, 113, 114, 115. The first addend is 115. Now I can count to find the second addend: 10, 20, 30, 40, 50, 60, 70, 71, 72, 73, 74, 75, 76. The second addend is 76.

_1_ hundreds   _9_ tens   _1_ ones

1▮▮▮ = _____

_115_ + _76_ = _191_

Lesson 10: Use math drawings to represent the composition when adding a two-digit to a three-digit addend.

45

> I can use $87 - 7$ to help me solve $87 - 8$. Since the difference in the first problem is 80, the difference in the second problem must be 1 less than 80 because I am only subtracting 1 more.

1. Solve using mental math.

   $7 - 6 = \underline{\textbf{1}}$ $87 - 6 = \underline{\textbf{81}}$ $87 - 7 = \underline{\textbf{80}}$ $87 - 8 = \underline{\textbf{79}}$

2. Solve using your place value chart and place value disks. Unbundle a ten if needed. Think about which problems you can solve mentally, too!

   a. $28 - 7 = \underline{\textbf{21}}$

   > I can solve this one mentally! I can subtract 7 ones from 8 ones. That leaves 2 tens 1 one, 21.

   b. $28 - 9 = \underline{\textbf{19}}$

   > I can use my chart and place value disks to solve this problem.

   > I only need to show 28 because I'm taking a part, 9, from the whole, 28.

   > I can't subtract 9 ones from 8 ones, so I change 1 ten for 10 ones. Now I have 1 ten 18 ones, so I can subtract 9 ones.

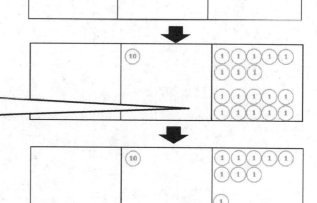

   > So $28 - 9 = 19$.

EUREKA MATH

Lesson 11:   Represent subtraction with and without the decomposition of 1 ten as 10 ones with manipulatives.

© 2018 Great Minds®. eureka-math.org

49

3. Solve 56 — 28, and explain your strategy.

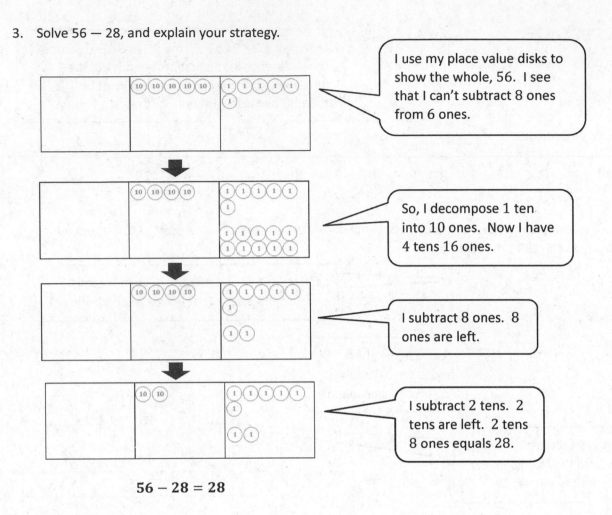

I use my place value disks to show the whole, 56. I see that I can't subtract 8 ones from 6 ones.

So, I decompose 1 ten into 10 ones. Now I have 4 tens 16 ones.

I subtract 8 ones. 8 ones are left.

I subtract 2 tens. 2 tens are left. 2 tens 8 ones equals 28.

$$56 - 28 = 28$$

4. The number of marbles in this jar is marked on the front. Miss Clark took 26 marbles out of the jar. How many marbles are left? Complete the number sentence to find out.

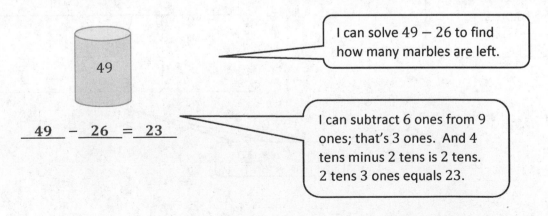

I can solve 49 — 26 to find how many marbles are left.

I can subtract 6 ones from 9 ones; that's 3 ones. And 4 tens minus 2 tens is 2 tens. 2 tens 3 ones equals 23.

$$\underline{\ 49\ } - \underline{\ 26\ } = \underline{\ 23\ }$$

Lesson 11:   Represent subtraction with and without the decomposition of 1 ten as 10 ones with manipulatives.

EUREKA MATH

Name _____   Date _____

1. Solve using mental math.

a. 6 - 5 = _____   26 - 5 = _____   26 - 6 = _____   26 - 7 = _____

b. 8 - 7 = _____   58 - 7 = _____   58 - 8 = _____   58 - 9 = _____

2. Solve using your place value chart and place value disks. Unbundle a ten, if needed.
    Think about which problems you can solve mentally, too!

a. 36 - 5 = _____   36 - 7 = _____

b. 37 - 6 = _____   37 - 8 = _____

c. 40 - 5 = _____   41 - 5 = _____

d. 58 - 32 =   58 - 29 = _____

e. 60 - 26 = _____   62 - 26 = _____

f. 70 - 41 = _____   80 - 41 = _____

EUREKA
MATH

Lesson 11:   Represent subtraction with and without the decomposition of
               1 ten as 10 ones with manipulatives.

© 2018 Great Minds®. eureka-math.org

51

3. Solve and explain your strategy.

a.

41 - 27 = _____

b.

67 - 28 = _____

4. The number of marbles in each jar is marked on the front. Miss Clark took 37 marbles out of each jar. How many marbles are left in each jar? Complete the number sentence to find out.

a. _____ - _____ = _____   b. _____ - _____ = _____

c. _____ - _____ = _____   d. _____ - _____ = _____

52   Lesson 11:   Represent subtraction with and without the decomposition of
1 ten as 10 ones with manipulatives.

© 2018 Great Minds®. eureka-math.org

EUREKA
MATH

1. Use place value disks to solve the problem.
   Rewrite the problem vertically, and record each step.

   71 − 27

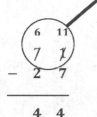

4 4

> I show the whole, 71, with my place value disks. I don't show 27 because it's already inside 71. When I subtract the part I know, 27, I'll find the missing part.

> I rewrite the problem in vertical form. Like a detective, I have to look carefully at the whole when subtracting, so I draw a magnifying glass around 71 to see if I need to do any unbundling.

> What I do with the disks, I need to do in vertical form.

> I can't subtract 7 ones from 1 one, so I need to decompose, or unbundle, a ten.

> Now I have 6 tens 11 ones. I'm ready to subtract!

> 11 ones − 7 ones = 4 ones.
> 6 tens − 2 tens = 4 tens.
> 4 tens 4 ones is 44.

2. Some Grade 1 and Grade 2 students voted on their favorite fruit. The table shows the number of votes for each fruit.

Types of Fruit	Number of Votes
Pineapple	26
Mango	18
Apple	15
Orange	35
Peach	43

a. How many more students voted for orange than pineapple? Show your work.

> The tape diagram helps me see that I'm looking for the difference between 35 and 26. I can subtract using the vertical form to find the answer.

> I can't take 6 ones from 5 ones, so I unbundle 1 ten. Now I have 2 tens 15 ones.
>
> 15 ones − 6 ones = 9 ones.

O | 35

P | 26

?

$$35 - 26 = ?$$

$$\begin{array}{r} \overset{2}{\cancel{3}}\ \overset{15}{\cancel{5}} \\ -\ 2\ \ 6 \\ \hline 9 \end{array}$$

9 more students voted for orange than for pineapple.

b. How many fewer students voted for mango than for pineapple? Show your work.

> The tape diagram helps me see that I'm looking for the difference between 18 and 26.

?

M | 18

P | 26

$$\begin{array}{r} \overset{1}{\cancel{2}}\ \overset{16}{\cancel{6}} \\ -\ 1\ \ 8 \\ \hline 8 \end{array}$$

$$26 - 18 = ?$$

8 fewer students voted for mango than for pineapple.

EUREKA MATH®

Name _____   Date _____

1. Use place value disks to solve each problem. Rewrite the problem vertically, and record each step as shown in the example.

   a. 34-18

   ```
      2  14
      3̶4̶
   -  1 8
   ─────
       1 6
   ```

   b. 41-16

   c. 33-15

   d. 46-18

   e. 62-27

   f. 81-34

2. Some first- and second-grade students voted on their favorite drink. The table shows the number of votes for each drink.

Types of Drink	Number of Votes
Milk	28
Apple Juice	19
Grape Juice	16
Fruit Punch	37
Orange Juice	44

a. How many more students voted for fruit punch than for milk? Show your work.

b. How many more students voted for orange juice than for grape juice? Show your work.

c. How many fewer students voted for apple juice than for milk? Show your work.

Lesson 12: Relate manipulative representations to a written method.

EUREKA MATH

1. Solve by writing the problem vertically. Check your result by drawing chips on the place value chart. Change 1 ten for 10 ones, when needed.

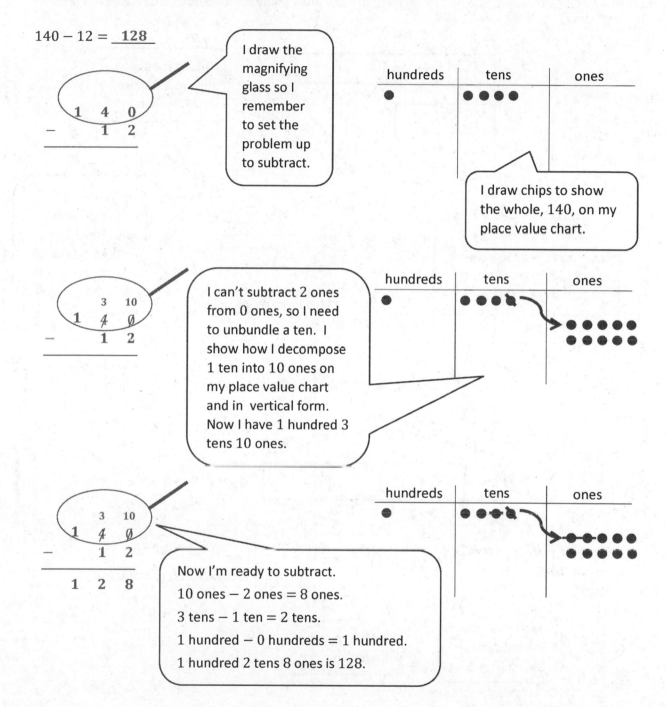

$140 - 12 = \underline{\ 128\ }$

I draw the magnifying glass so I remember to set the problem up to subtract.

I draw chips to show the whole, 140, on my place value chart.

I can't subtract 2 ones from 0 ones, so I need to unbundle a ten. I show how I decompose 1 ten into 10 ones on my place value chart and in vertical form. Now I have 1 hundred 3 tens 10 ones.

Now I'm ready to subtract.

10 ones − 2 ones = 8 ones.

3 tens − 1 ten = 2 tens.

1 hundred − 0 hundreds = 1 hundred.

1 hundred 2 tens 8 ones is 128.

EUREKA MATH

Lesson 14:   Represent subtraction with and without the decomposition when there is a three-digit minuend.

© 2018 Great Minds®. eureka-math.org

61

2. Solve and show your work. Draw a place value chart and chips, if needed.

a. Ana has 173 marbles. She has 27 more than Rico. How many marbles does Rico have?

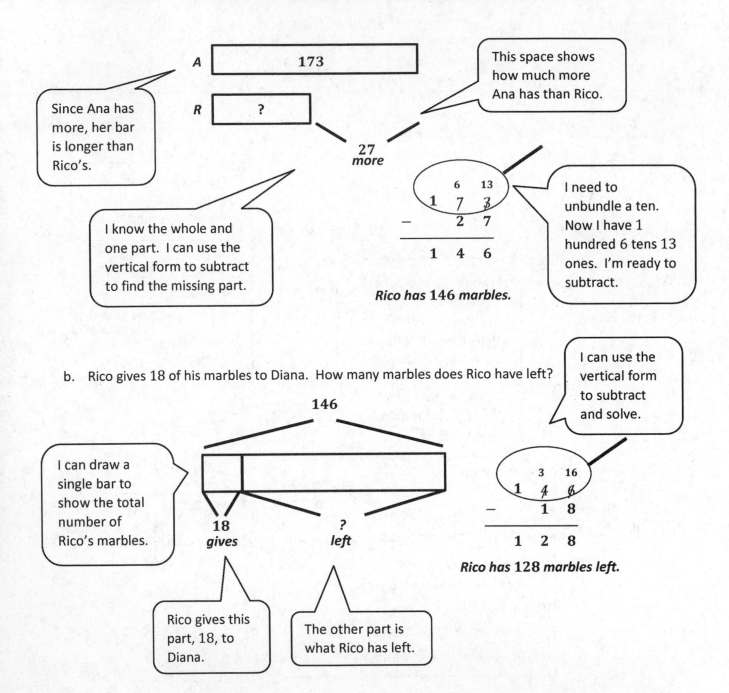

Since Ana has more, her bar is longer than Rico's.

This space shows how much more Ana has than Rico.

I know the whole and one part. I can use the vertical form to subtract to find the missing part.

I need to unbundle a ten. Now I have 1 hundred 6 tens 13 ones. I'm ready to subtract.

$$
\begin{array}{r}
1\ \ {}^{6}7\ \ {}^{13}\cancel{3} \\
-\ \ \ \ 2\ \ 7 \\
\hline
1\ \ 4\ \ 6
\end{array}
$$

Rico has **146 marbles.**

b. Rico gives 18 of his marbles to Diana. How many marbles does Rico have left?

I can use the vertical form to subtract and solve.

I can draw a single bar to show the total number of Rico's marbles.

146

18
gives

?
left

Rico gives this part, 18, to Diana.

The other part is what Rico has left.

$$
\begin{array}{r}
1\ \ {}^{3}\cancel{4}\ \ {}^{16}\cancel{6} \\
-\ \ \ \ 1\ \ 8 \\
\hline
1\ \ 2\ \ 8
\end{array}
$$

Rico has **128 marbles left.**

Lesson 14: Represent subtraction with and without the decomposition when there is a three-digit minuend.

EUREKA MATH

1. Solve using the vertical form. Show the subtraction on the place value chart with chips. Exchange 1 ten for 10 ones, if necessary.

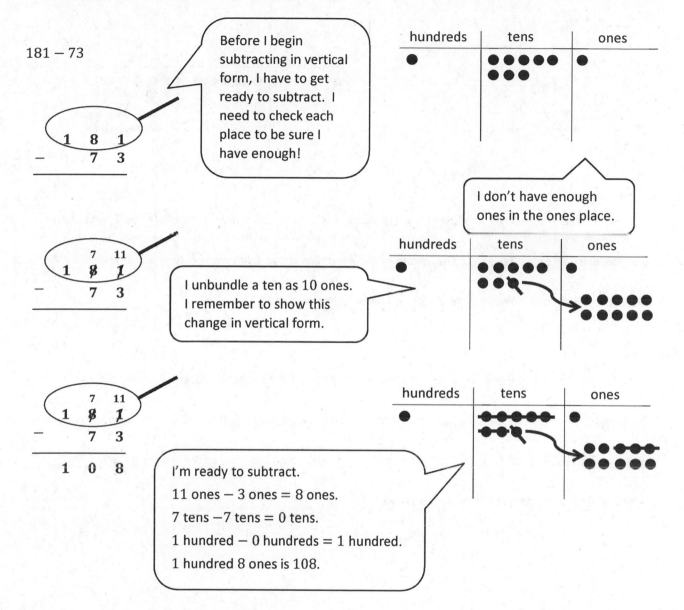

181 − 73

Before I begin subtracting in vertical form, I have to get ready to subtract. I need to check each place to be sure I have enough!

I don't have enough ones in the ones place.

I unbundle a ten as 10 ones. I remember to show this change in vertical form.

I'm ready to subtract.

11 ones − 3 ones = 8 ones.

7 tens − 7 tens = 0 tens.

1 hundred − 0 hundreds = 1 hundred.

1 hundred 8 ones is 108.

EUREKA MATH

Lesson 15:   Represent subtraction with and without the decomposition when there is a three-digit minuend.

65

© 2018 Great Minds®. eureka-math.org

2. Maya solved 157 – 39 vertically and on her place value chart. Explain what Maya did correctly and what she needs to fix.

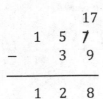

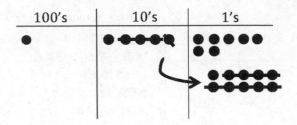

a. Maya correctly *models the problem on the place value chart. She shows the whole, 157, and then she decomposes 1 ten as 10 ones. She changes the model to show 1 hundred 4 tens 17 ones. After she crosses off 3 tens 9 ones, the model shows the correct answer, 118.*

b. Maya needs to fix *the vertical form. She forgot to draw the magnifying glass, which would have reminded her to look carefully to set the problem up for subtraction. She didn't show the change in the tens place, so she subtracted 3 tens from 5 tens, instead of subtracting from 4 tens. That's why she got the wrong answer, 128.*

     Lesson 15:   Represent subtraction with and without the decomposition when there is a three-digit minuend.

EUREKA MATH

© 2018 Great Minds®. eureka-math.org

Solve the following word problems. Use the RDW process.

1. Audrey put 56 beads on a necklace. Some beads fell off, but she still has 28 left. How many beads fell off?

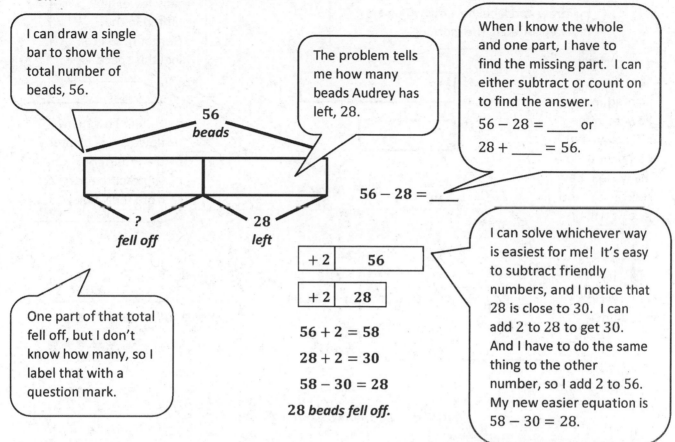

> I can draw a single bar to show the total number of beads, 56.

> The problem tells me how many beads Audrey has left, 28.

> When I know the whole and one part, I have to find the missing part. I can either subtract or count on to find the answer.
> $56 - 28 =$ _____ or
> $28 +$ _____ $= 56$.

56
beads

$56 - 28 =$ ___

?
fell off

28
left

> One part of that total fell off, but I don't know how many, so I label that with a question mark.

| + 2 | 56 |

| + 2 | 28 |

$56 + 2 = 58$

$28 + 2 = 30$

$58 - 30 = 28$

28 beads fell off.

> I can solve whichever way is easiest for me! It's easy to subtract friendly numbers, and I notice that 28 is close to 30. I can add 2 to 28 to get 30. And I have to do the same thing to the other number, so I add 2 to 56. My new easier equation is $58 - 30 = 28$.

2. Farmer Ben picks 87 apples. 26 apples are green, 20 are yellow, and the rest are red. How many apples are red?

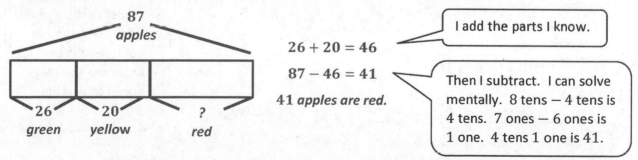

87
apples

26
green

20
yellow

?
red

$26 + 20 = 46$

$87 - 46 = 41$

41 apples are red.

> I add the parts I know.

> Then I subtract. I can solve mentally. 8 tens − 4 tens is 4 tens. 7 ones − 6 ones is 1 one. 4 tens 1 one is 41.

3. Ava planted 45 flowers in the morning. She planted 26 fewer flowers in the afternoon. How many flowers did she plant altogether?

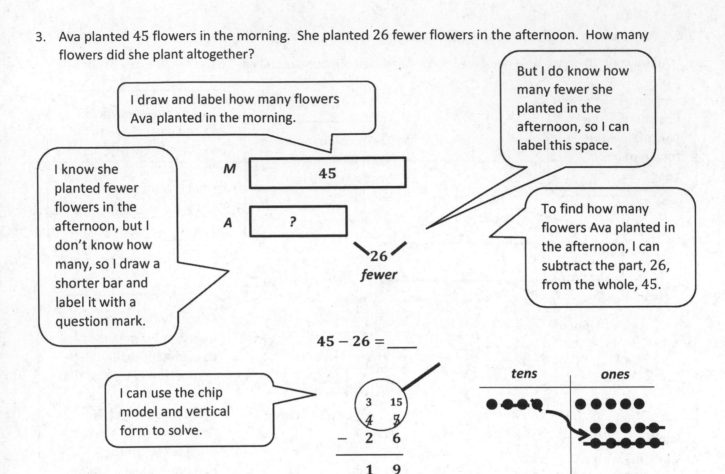

I draw and label how many flowers Ava planted in the morning.

But I do know how many fewer she planted in the afternoon, so I can label this space.

I know she planted fewer flowers in the afternoon, but I don't know how many, so I draw a shorter bar and label it with a question mark.

To find how many flowers Ava planted in the afternoon, I can subtract the part, 26, from the whole, 45.

M   45

A   ?

26 fewer

$$45 - 26 = \underline{}$$

I can use the chip model and vertical form to solve.

tens   ones

Ava planted **19** *flowers in the afternoon.*

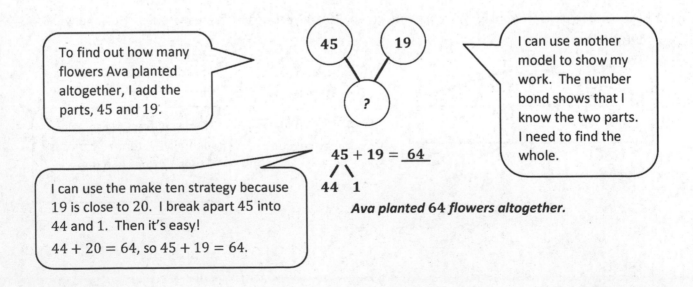

To find out how many flowers Ava planted altogether, I add the parts, 45 and 19.

45   19

?

I can use another model to show my work. The number bond shows that I know the two parts. I need to find the whole.

$$45 + 19 = \underline{\ 64\ }$$

44 1

I can use the make ten strategy because 19 is close to 20. I break apart 45 into 44 and 1. Then it's easy!

$44 + 20 = 64$, so $45 + 19 = 64$.

Ava planted **64** *flowers altogether.*

Lesson 16:   Solve one- and two-step word problems within 100 using strategies based on place value.

EUREKA MATH

Name _____ Date _____

Solve the following word problems. Use the RDW process.

1. Vicki modeled the following problem with a tape diagram.

   Eighty-two students are in the math club. 35 students are in the science club. How many more students are in the math club than science club?

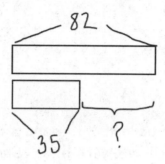

   Show another model to solve the problem. Write your answer in a sentence.

Lesson 16:   Solve one- and two-step word problems within 100 using strategies based on place value.

© 2018 Great Minds®. eureka-math.org

71

2. Forty-six birds sat on a wire. Some flew away, but 29 stayed. How many birds flew away? Show your work.

3. Ian bought a pack of 47 water balloons. 19 were red, 16 were yellow, and the rest were blue. How many water balloons were blue? Show your work.

4. Daniel read 54 pages of his book in the morning. He read 27 fewer pages in the afternoon. How many pages did Daniel read altogether? Show your work.

Lesson 16:   Solve one- and two-step word problems within 100 using strategies based on place value.

EUREKA MATH

1. Solve mentally.

> I have to pay attention to the unit. 8 tens equal 80. 1 hundred more than 80 is 180.

1 ten more than 8 ones = __18__

1 hundred more than 8 ones = __108__

1 hundred more than 8 tens = __180__

$10 + 8 =$ __18__

$100 + 8 =$ __108__

$100 + 80 =$ __180__

12 ones + 2 ones = __1__ ten(s) __4__ one(s)

12 tens + 2 tens = __1__ hundred(s) __4__ tens(s)

$12 + 2 =$ __14__

$120 + 20 =$ __140__

> Both of these number sentences show $12 + 2$. What's different is the unit. Adding 10 ones makes 1 ten. Adding 10 tens makes 1 hundred.

2. Solve.

7 ones + 8 ones = __1__ ten __5__ ones

7 tens + 8 tens = __1__ hundred __5__ tens

$7 + 8 =$ __15__

$70 + 80 =$ __150__

> 15 tens is the same as 150. I can show 15 tens on a place value chart. When I circle 10 tens, I make a hundred, and there are 5 tens left.

3. Fill in the blanks. Then, complete the addition sentence.

$$54 \xrightarrow{+6} \underline{\ 60\ } \xrightarrow{+40} \underline{\ 100\ } \xrightarrow{+10} \underline{\ 110\ } \xrightarrow{+10} \underline{\ 120\ } \xrightarrow{+100} \underline{\ 220\ }$$

$54 +$ __166__ $=$ __220__

> First, I add 6 to make a ten, 60. Then I add 40 to make a hundred. It's easy to add on 2 more tens and a hundred to make 220.

Lesson 17:   Use mental strategies to relate compositions of 10 tens as
                   1 hundred to 10 ones as 1 ten.

73

© 2018 Great Minds®. eureka-math.org

Name _____   Date _____

1. Solve mentally.

    a.   4 ones + _____ = 1 ten   4 + _____ = 10

         4 tens + _____ = 1 hundred   40 + _____ = 100

    b.   1 ten = _____ + 7 ones   10 = _____ + 7

         1 hundred = _____ + 7 tens   100 = _____ + 70

    c.   1 ten more than 9 ones = _____   10 + 9 = _____

         1 hundred more than 9 ones = _____   100 + 9 = _____

         1 hundred more than 9 tens = _____   100 + 90 = _____

    d.   2 ones + 8 ones = _____ ten   2 + 8 = _____

         2 tens + 8 tens = _____ hundred   20 + 80 = _____

    e.   5 ones + 6 ones = ____ ten(s) ____ one(s)   5 + 6 = _____

         5 tens + 6 tens = ____ hundred(s) ____ ten(s)   50 + 60 = _____

    f.   14 ones + 4 ones = _____ ten(s) _____ one(s)   14 + 4 = _____

         14 tens + 4 tens = _____ hundred(s) _____ tens(s)   140 + 40 = _____

2. Solve.

a. 6 ones + 5 ones = _____ ten _____ one   6 + 5 = _____

    6 tens + 5 tens = _____ hundred _____ ten   60 + 50 = _____

b. 5 ones + 7 ones = _____ ten _____ ones   5 + 7 = _____

    5 tens + 7 tens = _____ hundred _____ tens   50 + 70 = _____

c. 9 ones + 8 ones = _____ ten _____ ones   9 + 8 = _____

    9 tens + 8 tens = _____ hundred _____ tens   90 + 80 = _____

3. Fill in the blanks. Then, complete the addition sentence. The first one is done for you.

a. $36 \xrightarrow{+4} \underline{40} \xrightarrow{+60} \underline{100} \xrightarrow{+30} \underline{130}$ b. $78 \xrightarrow{+2} \underline{\quad} \xrightarrow{+10} \underline{\quad} \xrightarrow{+10} \underline{\quad}$

    $36 + \underline{94} = \underline{130}$ $78 + \underline{\quad} = \underline{\quad}$

c. $61 \xrightarrow{+9} \underline{\quad} \xrightarrow{+10} \underline{\quad} \xrightarrow{+10} \underline{\quad} \xrightarrow{+10} \underline{\quad} \xrightarrow{+100} \underline{\quad}$

    $61 + \underline{\quad} = \underline{\quad}$

d. $27 \xrightarrow{+3} \underline{\quad} \xrightarrow{+70} \underline{\quad} \xrightarrow{+100} \underline{\quad}$

    $27 + \underline{\quad} = \underline{\quad}$

Lesson 17:   Use mental strategies to relate compositions of 10 tens as
                    1 hundred to 10 ones as 1 ten.

EUREKA
MATH®

1. Solve using your place value chart and place value disks.

    35 + 76 = __111__   36 + 86 = __122__

    These problems are very similar. Just from looking at the tens and ones, I know that my second answer will have 1 more ten and 1 more one than the first answer.

    36 is one more than 35, and 86 is 10 more than 76.

2. Circle the statements that are true as you solve the problem using place value disks.

    136 + 58

    (I change 10 ones for 1 ten.)

    I change 10 tens for 1 hundred.

    The total of the two parts is 184.

    (The total of the two parts is 194.)

    I can set up this problem with place value disks and add like units. 6 ones and 8 ones are 14 ones. I can change 10 ones for 1 ten. I'll have 4 ones left over. Then, 3 tens + 5 tens + 1 ten equals 9 tens. 1 hundred + 9 tens + 4 ones = 194.

3. Solve the problem using your place value disks, and fill in the missing total. Then, write an addition sentence that relates to the number bond.

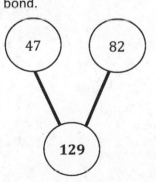

   I can change 10 tens for 1 hundred!

   Now I have 1 hundred 2 tens 9 ones, 129.

   I have 9 ones. I can't make a ten.

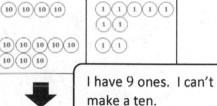

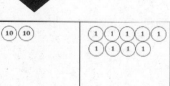

   Addition sentence:
   47 + 82 = 129

1. Solve the following problems using the vertical form, your place value chart, and place value disks. Bundle a ten or hundred, if needed.

> I can solve this one mentally! 69 is close to 70, so I can think $24 + 70 = 94$. Then, I can just subtract 1, and the answer is 93.

a. $24 + 69 =$ __93__

b. $137 + 63 =$ **200**

> I can use my chart and place value disks to solve.

> I write it in vertical form as I model it with my place value disks.

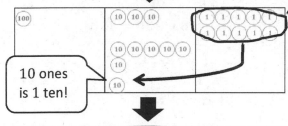

$$\begin{array}{r} 1\ \ 3\ \ 7 \\ +\ \ \ \ 6\ \ 3 \\ \hline \end{array}$$

> I bundle 10 ones and make a ten!

> I show the ten using new groups below, on the line below the tens place.

> 10 ones is 1 ten!

$$\begin{array}{r} 1\ \ 3\ \ 7 \\ +\ \ \ \ 6\ \ 3 \\ \hline {\scriptstyle 1}\ \ \ \\ 0 \end{array}$$

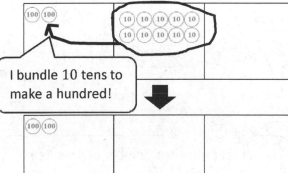

> I bundle 10 tens to make a hundred!

> Now I add the tens. 3 tens plus 6 tens plus 1 ten is 10 tens. I can bundle again to make 1 hundred! I show the hundred using new groups below again.

$$\begin{array}{r} 1\ \ 3\ \ 7 \\ +\ \ \ \ 6\ \ 3 \\ \hline {\scriptstyle 1}\ {\scriptstyle 1}\ \ \\ 0\ \ 0 \end{array}$$

$$\begin{array}{r} 1\ \ 3\ \ 7 \\ +\ \ \ \ 6\ \ 3 \\ \hline {\scriptstyle 1}\ {\scriptstyle 1}\ \ \\ 2\ \ 0\ \ 0 \end{array}$$

> Last, I add the hundreds. There are 2 hundreds.

2. Eighty-four girls attended swim school. Twenty-nine more boys attended than girls.

a. How many boys attended swim school?

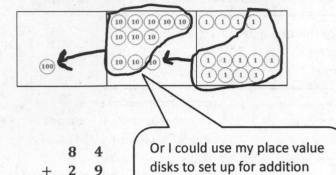

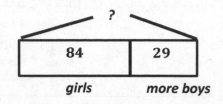

```
    8   4
+   2   9
  ̶1̶   ̶1̶
    1   1   3
```

Or I could use my place value disks to set up for addition with renaming. I can show 84 and 29 with disks and solve vertically.

I can draw a tape diagram to represent the story. I can use the make ten strategy to solve! (See below.)

$84 + 29 = \underline{\quad}$

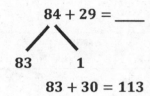

$83 + 30 = 113$

113 boys attended swim school.

b. How many boys and girls attended swim school?

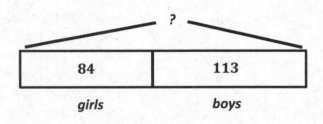

```
    1   1   3
+       8   4
    1   9   7
```

197 boys and girls attended swim school.

Now that I know the number of boys, I can add the girls and boys together to find the total. I can show my work using the vertical method.

EUREKA
MATH®

Solve vertically. Draw chips on the place value chart and bundle, when needed.

1. $58 + 74 =$ __132__

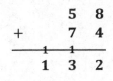

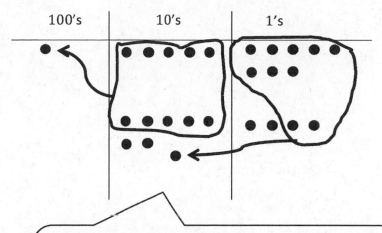

I show each step I make with chips vertically using new groups below.

I draw chips to show each addend. 8 ones plus 4 ones is 12 ones, or 1 ten 2 ones. I bundle 10 ones to make 1 ten. Now I add the tens. 5 tens plus 7 tens plus 1 more ten is 13 tens. I can bundle again! 10 tens makes 1 hundred. So, 13 tens is 1 hundred 3 tens.

2. For the box below, find and circle two numbers that add up to 160.

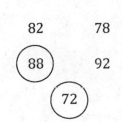

If I add 88 and 72, I can add 8 ones and 2 ones, which is 10 ones. I can bundle ten ones to make 1 ten! Then, I can add 8 tens plus 7 tens plus 1 ten to get 16 tens, or 160.

I see the trap; if I forgot to add another ten, I might have chosen 88 and 82 or 78 and 92.

EUREKA
MATH®

Lesson 20:   Use math drawings to represent additions with up to two
compositions and relate drawings to a written method.

85

© 2018 Great Minds®. eureka-math.org

Solve vertically. Draw chips on the place value chart and bundle, when needed.

1. $138 + 62 =$ __200__

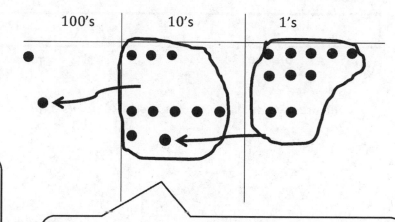

	1	3	8
+		6	2
	1	1	
	2	0	0

My model matches the vertical method. I bundled twice, and I can show the new units with new groups below.

Renaming the tens is just like renaming ones. I have to look for 10 of a unit to make the next higher value unit. So, 10 ones make 1 ten, and 10 tens make 1 hundred!

2. The orange team scored 26 fewer points than the green team. The orange team scored 49 points.

   a. How many points did the green team score?

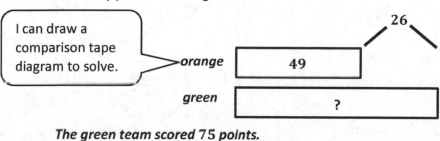

I can draw a comparison tape diagram to solve.

$49 + 26 =$ __?__

$50 + 26 = 76$

$76 - 1 = 75$

The green team scored 75 points.

I don't need to solve with chips because 49 is close to 50. I can add 50 and 26, which makes 76. Then, I can subtract 1 since 49 is 1 less than 50. I can use the same strategy for Part (b).

EUREKA MATH

Lesson 21:   Use math drawings to represent additions with up to two compositions and relate drawings to a written method.

89

© 2018 Great Minds®. eureka-math.org

b. How many points did the orange and green teams score altogether?

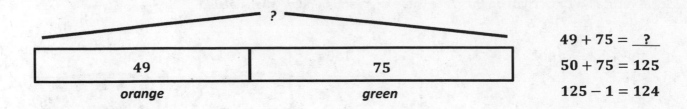

$49 + 75 = \underline{\ ?\ }$

$50 + 75 = 125$

$125 - 1 = 124$

The orange and green team scored 124 points altogether.

Lesson 21: Use math drawings to represent additions with up to two compositions and relate drawings to a written method.

EUREKA MATH

1. Look to make 10 ones or 10 tens to solve the following problems using place value strategies.

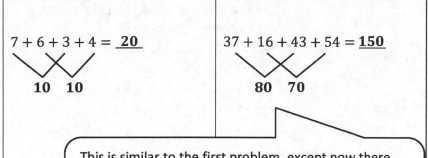

$7 + 6 + 3 + 4 =$ __20__

10   10

$37 + 16 + 43 + 54 =$ __150__

80   70

$86 + 34 + 33 + 67 =$ __220__

120   100

This is similar to the first problem, except now there are tens. When I add 37 plus 43, I know 7 ones plus 3 ones equals 10 ones, or 1 ten. Then, 3 tens plus 4 tens equals 7 tens. 7 tens + 1 ten = 8 tens, or 80.

I can group 86 and 34 together because 6 and 4 make 10. 8 tens plus 3 tens equals 11 tens. When I add 1 more ten, I get 12 tens, which is 120. $120 + 100 = 220$.

2. The table shows the top five soccer teams and their total points scored this season.

Teams	Points
Red	48
Yellow	39
Green	52
Blue	41
Orange	42

a. How many points did the yellow, orange, and blue teams score together?

$39 + 42 + 41 =$ __122__

$80 + 42$

Since 9 and 1 make ten, I added 39 and 41 first. I know that $30 + 40 = 70$, and $70 + 10 = 80$. Then, $80 + 42 = 122$.

The yellow, orange, and blue teams scored 122 points.

EUREKA MATH   Lesson 22:   Solve additions with up to four addends with totals within 200 with   93
                                 and without two compositions of larger units.

© 2018 Great Minds®. eureka-math.org

b. Which two teams scored a total of 90 points?

$$48 + 42 = 90$$

The red and orange teams scored 90 points.

> I can look for a total of 9 tens. 4 tens plus 4 tens is 8 tens, which is only 80. But, don't forget the ones! 8 ones plus 2 ones equals 10 ones, or 1 ten. So 8 tens and 1 more ten is 9 tens, or 90.

Lesson 22: Solve additions with up to four addends with totals within 200 with and without two compositions of larger units.

EUREKA MATH

Name _____   Date _____

1. Look to make 10 ones or 10 tens to solve the following problems using place value strategies.

a.		
6 + 3 + 7= _____	36 + 23 + 17= _____	126 + 23 + 17= _____
b.		
8 + 2 + 5 = _____	38 + 22 + 75 = _____	18 + 62 + 85 = _____
c.		
9 + 4 + 1 + 6 = _____	29 + 34 + 41 + 16 = _____	81 + 34 + 19 + 56 = _____

Lesson 22:   Solve additions with up to four addends with totals within 200 with and without two compositions of larger units.

95

2. The table shows the top six soccer teams and their total points scored this season.

Teams	Points
Red	29
Yellow	38
Green	41
Blue	76
Orange	52
Black	24

a. How many points did the yellow and orange teams score together?

b. How many points did the yellow, orange, and blue teams score together?

c. How many points did the red, green, and black teams score together?

d. Which two teams scored a total of 70 points?

e. Which two teams scored a total of 100 points?

Lesson 22:   Solve additions with up to four addends with totals within 200 with
and without two compositions of larger units.

© 2018 Great Minds®. eureka-math.org

EUREKA
MATH

1. Solve using number bonds to subtract from 100.

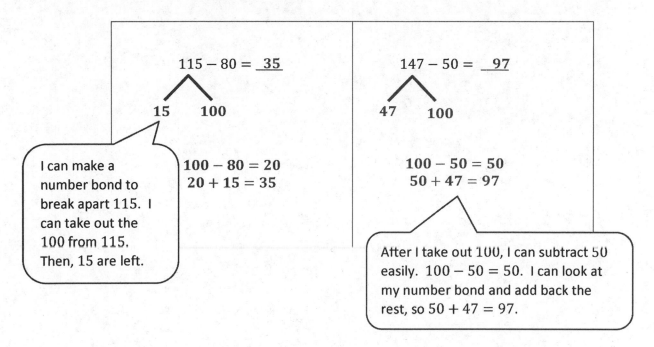

$$115 - 80 = \underline{\ 35\ }$$

15   100

I can make a number bond to break apart 115. I can take out the 100 from 115. Then, 15 are left.

$$100 - 80 = 20$$
$$20 + 15 = 35$$

$$147 - 50 = \underline{\ 97\ }$$

47   100

$$100 - 50 = 50$$
$$50 + 47 = 97$$

After I take out 100, I can subtract 50 easily. $100 - 50 = 50$. I can look at my number bond and add back the rest, so $50 + 47 = 97$.

2. Jana sold 60 fewer candles than Charlotte. Charlotte sold 132 candles. How many candles did Jana sell? Solve using a number bond.

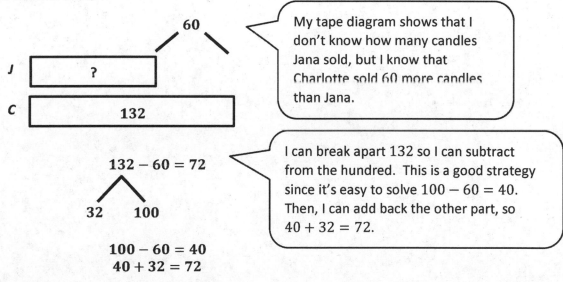

60

J   ?

C   132

My tape diagram shows that I don't know how many candles Jana sold, but I know that Charlotte sold 60 more candles than Jana.

$$132 - 60 = 72$$

32   100

I can break apart 132 so I can subtract from the hundred. This is a good strategy since it's easy to solve $100 - 60 = 40$. Then, I can add back the other part, so $40 + 32 = 72$.

$$100 - 60 = 40$$
$$40 + 32 = 72$$

Jana sold 72 candles.

Lesson 23:   Use number bonds to break apart three-digit minuends and subtract from the hundred.

97

EUREKA MATH

© 2018 Great Minds®. eureka-math.org

Name _____   Date _____

1. Solve using number bonds to subtract from 100. The first one has been done for you.

a. 105 – 90 = 15	b. 121 – 90
100 5 100 – 90 = 10 10 + 5 = 15	
c. 112 – 80	d. 135 – 70
e. 136 – 60	f. 129 – 50

EUREKA
MATH

Lesson 23:   Use number bonds to break apart three-digit minuends and subtract from
                the hundred.

© 2018 Great Minds®. eureka-math.org

99

g. 156 – 80	h. 138 – 40

2. Monica incorrectly solved 132 - 70 to get 102. Show her how to solve it correctly.

Monica's work:	Correct way to solve 132 – 70:
132 - 70 = _____ 100 32 100 - 30 = 70 70 + 32 = 102	

3. Billy sold 50 fewer magazines than Alex. Alex sold 128 magazines. How many magazines did Billy sell? Solve using a number bond.

Lesson 23:   Use number bonds to break apart three-digit minuends and subtract from the hundred.

EUREKA MATH

1. Solve using mental math. If you cannot solve mentally, use your place value chart and place value disks.

$47 - 7 =$ __40__ $47 - 8 =$ __39__ $147 - 47 =$ __100__ $147 - 48 =$ __99__

> I can use $147 - 47$ to help me solve $147 - 48$. Since the difference in the first problem is 100, the difference in the second problem must be 1 less than 100 because I am only subtracting 1 more.

2. Solve using your place value chart and place value disks. Unbundle the hundred or ten when necessary. Circle what you did to model each problem.

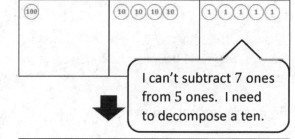

$145 - 87 =$ __58__

I unbundled the hundred.   (Yes) No

I unbundled a ten.   (Yes) No

> I can't subtract 7 ones from 5 ones. I need to decompose a ten.

> I only have 3 tens. That's not enough to subtract 8 tens! I need to unbundle the hundred.

> Now I have 15 ones. That's enough to subtract 7 ones.

> Now I have 13 tens and 15 ones. I am ready to subtract!
> 13 tens − 8 tens = 5 tens.
> 15 ones − 7 ones = 8 ones.
> 5 tens 8 ones is 58.

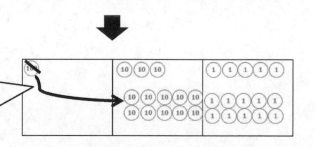

EUREKA MATH

Lesson 24:   Use manipulatives to represent subtraction with decompositions of 1 hundred as 10 tens and 1 ten as 10 ones.

101

© 2018 Great Minds®. eureka-math.org

3. 76 pencils in the basket are sharpened. The basket has 132 pencils. How many pencils are not sharpened?

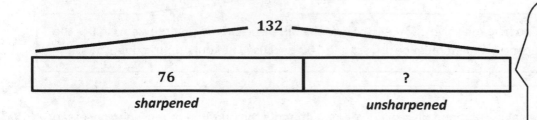

132

76	?
sharpened	*unsharpened*

$132 - 76 = ?$

> My tape diagram shows that 132 is the total. I know that one part is 76 sharpened pencils. I am solving for the number of pencils that are not sharpened. That's my unknown.

$$76 \xrightarrow{+4} 80 \xrightarrow{+20} 100 \xrightarrow{+32} 132$$

56 *pencils are not sharpened.*

> I can use the arrow way to find the missing part. I can start at 76 and add 4 to get to a friendly number, 80. Then, I can add 20 to get to 1 hundred. Then, 32 more is 132. So, $20 + 32 + 4 = 56$.

Lesson 24:   Use manipulatives to represent subtraction with decompositions of 1 hundred as 10 tens and 1 ten as 10 ones.

EUREKA MATH

Name _____   Date _____

1. Solve using mental math. If you cannot solve mentally, use your place value chart and place value disks.

   a. 38 – 8 = _____   38 – 9 = _____   138 – 38 = _____   138 – 39 = _____

   b. 130 – 20 = _____   130 – 30 = _____   130 – 40 = _____

2. Solve using your place value chart and place value disks. Unbundle the hundred or ten when necessary. Circle what you did to model each problem.

a.	b.
115 – 50 = _____	125 – 57 = _____
I unbundled the hundred. Yes No I unbundled a ten. Yes No	I unbundled the hundred. Yes No I unbundled a ten. Yes No
c.	d.
88 – 39 = _____	186 – 39 = _____
I unbundled the hundred. Yes No I unbundled a ten. Yes No	I unbundled the hundred. Yes No I unbundled a ten. Yes No
e.	f.
162 – 85 = _____	172 – 76 = _____
I unbundled the hundred. Yes No I unbundled a ten. Yes No	I unbundled the hundred. Yes No I unbundled a ten. Yes No

EUREKA MATH

Lesson 24: Use manipulatives to represent subtraction with decompositions of 1 hundred as 10 tens and 1 ten as 10 ones.

103

© 2018 Great Minds®. eureka-math.org

g. $121 - 89 = $ _____	h. $131 - 98 = $ _____
I unbundled the hundred. Yes No I unbundled a ten. Yes No	I unbundled the hundred. Yes No I unbundled a ten. Yes No
i. $140 - 65 = $ _____	j. $150 - 56 = $ _____
I unbundled the hundred. Yes No I unbundled a ten. Yes No	I unbundled the hundred. Yes No I unbundled a ten. Yes No
k. $163 - 78 = $ _____	l. $136 - 87 = $ _____
I unbundled the hundred. Yes No I unbundled a ten. Yes No	I unbundled the hundred. Yes No I unbundled a ten. Yes No

3. 96 crayons in the basket are broken. The basket has 182 crayons. How many crayons are not broken?

Lesson 24:   Use manipulatives to represent subtraction with decompositions of 1 hundred as 10 tens and 1 ten as 10 ones.

© 2018 Great Minds®. eureka-math.org

EUREKA MATH®

1. Solve the following problems using the vertical form, your place value chart, and place value disks. Unbundle a ten or hundred when necessary. Show your work for each problem.

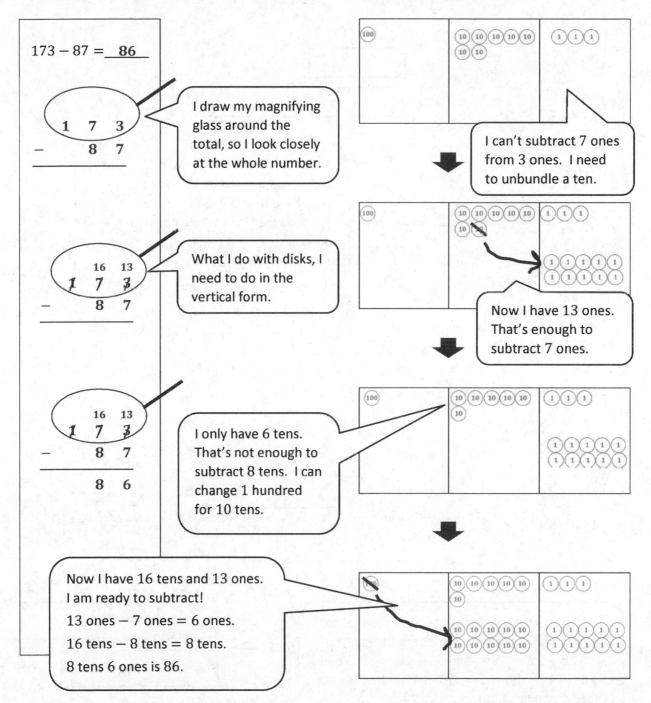

$173 - 87 = \underline{86}$

I draw my magnifying glass around the total, so I look closely at the whole number.

What I do with disks, I need to do in the vertical form.

I only have 6 tens. That's not enough to subtract 8 tens. I can change 1 hundred for 10 tens.

Now I have 16 tens and 13 ones. I am ready to subtract!
13 ones − 7 ones = 6 ones.
16 tens − 8 tens = 8 tens.
8 tens 6 ones is 86.

I can't subtract 7 ones from 3 ones. I need to unbundle a ten.

Now I have 13 ones. That's enough to subtract 7 ones.

2. Vazyl has $127. He has $65 more than Sergio. How much money does Sergio have?

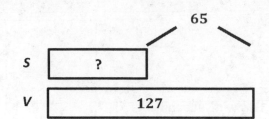

$127 - 65 = ?$

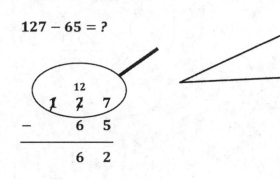

I can use the vertical method to figure out how much money Sergio has. I only have to unbundle the hundred because there are enough ones to subtract.

7 ones − 5 ones = 2 ones.

12 tens − 6 tens = 6 tens.

6 tens 2 ones is 62.

Sergio has 62 dollars.

3. Which problem will have the same answer as $122 - 66$? Show your work.

a. $144 - 55$

b. $126 - 62$

c. $166 - 22$

d. $144 - 88$

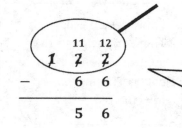

I can use the vertical form to solve $122 - 66$.

But I also know another strategy. If I add 22 to both numbers, the difference doesn't change. So, $122 + 22 = 144$. And $66 + 22 = 88$. That means $144 - 88 = 56$. I remember this; it's called compensation!

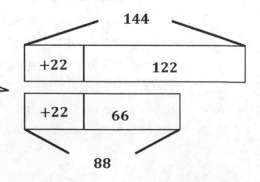

Lesson 25:   Relate manipulative representations to a written method.

EUREKA MATH

Name _____   Date _____

1. Solve the following problems using the vertical form, your place value chart, and place value disks. Unbundle a ten or hundred when necessary. Show your work for each problem.

a. 65 – 38	b. 66 – 49
c. 111 – 60	d. 120 – 67
e. 163 – 66	f. 184 – 95
g. 114 – 98	h. 154 – 85

2. Dominic has $167. He has $88 more than Mario. How much money does Mario have?

3. Which problem will have the same answer as 133 − 77? Show your work.

    a. 155 – 66

    b. 144 – 88

    c. 177 – 33

    d. 139 – 97

EUREKA
MATH

Solve vertically. Draw chips on the place value chart. Unbundle when needed.

$152 - 67 = \underline{85}$

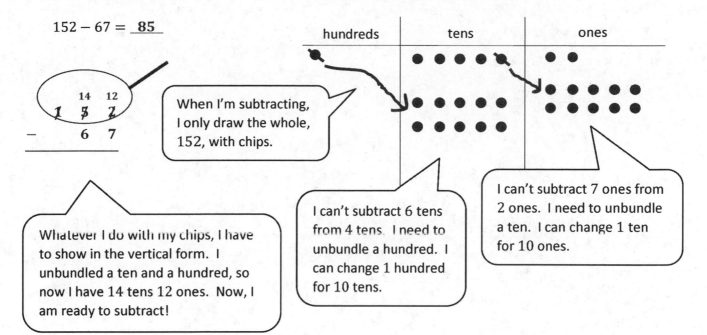

When I'm subtracting, I only draw the whole, 152, with chips.

I can't subtract 7 ones from 2 ones. I need to unbundle a ten. I can change 1 ten for 10 ones.

I can't subtract 6 tens from 4 tens. I need to unbundle a hundred. I can change 1 hundred for 10 tens.

Whatever I do with my chips, I have to show in the vertical form. I unbundled a ten and a hundred, so now I have 14 tens 12 ones. Now, I am ready to subtract!

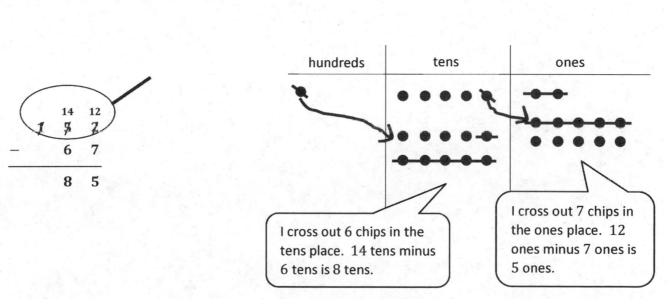

I cross out 6 chips in the tens place. 14 tens minus 6 tens is 8 tens.

I cross out 7 chips in the ones place. 12 ones minus 7 ones is 5 ones.

Lesson 26:   Use math drawings to represent subtraction with up to two decompositions and relate drawings to a written method.

109

EUREKA MATH

© 2018 Great Minds®. eureka-math.org

Solve vertically. Draw chips on the place value chart. Unbundle when needed.

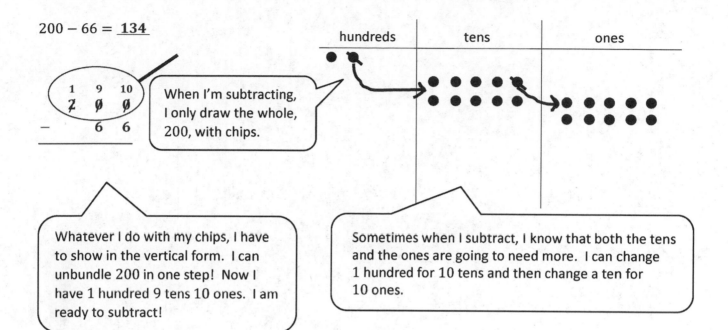

$200 - 66 = \underline{134}$

When I'm subtracting, I only draw the whole, 200, with chips.

Whatever I do with my chips, I have to show in the vertical form. I can unbundle 200 in one step! Now I have 1 hundred 9 tens 10 ones. I am ready to subtract!

Sometimes when I subtract, I know that both the tens and the ones are going to need more. I can change 1 hundred for 10 tens and then change a ten for 10 ones.

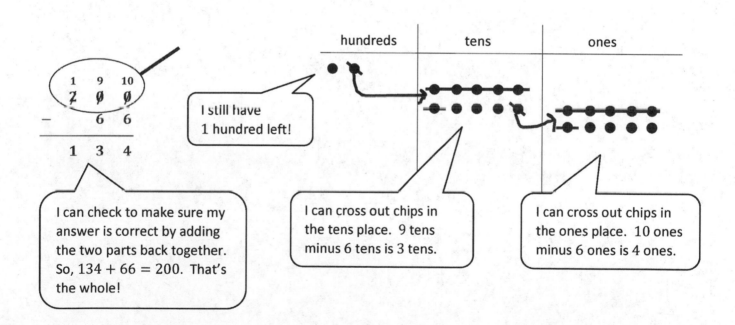

I still have 1 hundred left!

I can check to make sure my answer is correct by adding the two parts back together. So, $134 + 66 = 200$. That's the whole!

I can cross out chips in the tens place. 9 tens minus 6 tens is 3 tens.

I can cross out chips in the ones place. 10 ones minus 6 ones is 4 ones.

Name _____   Date _____

1. Solve vertically. Draw chips on the place value chart. Unbundle when needed.

a.   100 – 37 = _____

hundreds	tens	ones

b.   100 – 49 = _____

hundreds	tens	ones

c.   200 – 49 = _____

hundreds	tens	ones

EUREKA MATH®

Lesson 27:   Subtract from 200 and from numbers with zeros in the tens place.

115

© 2018 Great Minds®. eureka-math.org

d. 200 – 57 = _____

hundreds	tens	ones

e. 200 – 83 = _____

hundreds	tens	ones

2. Susan solved 200 – 91 and decided to add her *answer* to 91 to check her work. Explain why this strategy works.

Susan's work:	Explanation:
$$\begin{array}{r} 1\;9\;\;\; \\ 2\,0\,0^{10} \\ -\;\;9\,1 \\ \hline 1\,0\,9 \end{array}$$ $$\begin{array}{r} 1\,0\,9 \\ +\;\;9\,1 \\ \hline 2\,0\,0 \end{array}$$	_____ _____ _____ _____

EUREKA MATH

1. Solve vertically. Draw chips on the place value chart. Unbundle when needed.

$200 - 108 = \underline{92}$

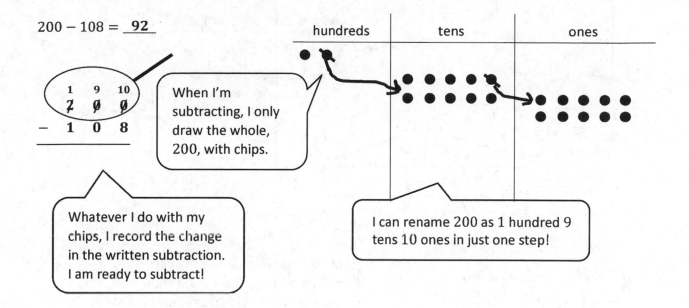

When I'm subtracting, I only draw the whole, 200, with chips.

Whatever I do with my chips, I record the change in the written subtraction. I am ready to subtract!

I can rename 200 as 1 hundred 9 tens 10 ones in just one step!

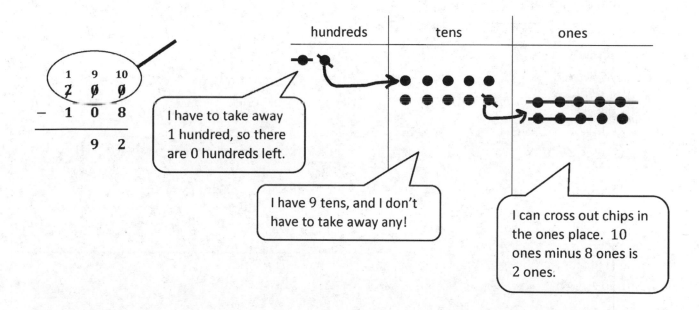

I have to take away 1 hundred, so there are 0 hundreds left.

I have 9 tens, and I don't have to take away any!

I can cross out chips in the ones place. 10 ones minus 8 ones is 2 ones.

EUREKA
MATH

Lesson 28:   Subtract from 200 and from numbers with zeros in the tens place.

117

© 2018 Great Minds®. eureka-math.org

2. Harry collected 200 baseball cards. He traded 127 of them and kept the rest. How many baseball cards did he keep?

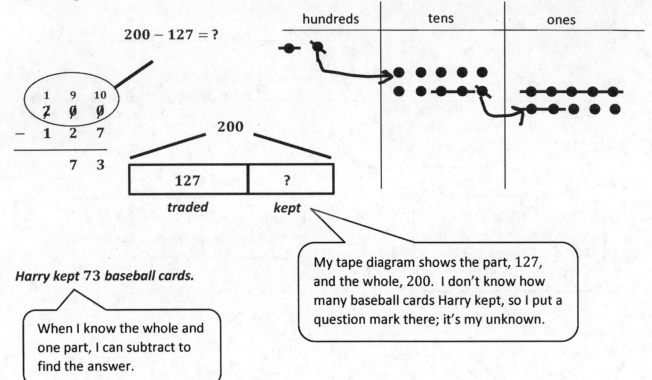

$$200 - 127 = ?$$

Harry kept 73 *baseball cards.*

When I know the whole and one part, I can subtract to find the answer.

My tape diagram shows the part, 127, and the whole, 200. I don't know how many baseball cards Harry kept, so I put a question mark there; it's my unknown.

Lesson 28:   Subtract from 200 and from numbers with zeros in the tens place.

EUREKA MATH

Name _____ Date _____

1. Solve vertically. Draw chips on the place value chart. Unbundle when needed.

a. 136 – 94 = _____

hundreds	tens	ones

b. 105 – 57 = _____

hundreds	tens	ones

c. 200 – 61 = _____

hundreds	tens	ones

EUREKA MATH

Lesson 28:   Subtract from 200 and from numbers with zeros in the tens place.

119

© 2018 Great Minds®. eureka-math.org

d. 200 – 107 = _____

hundreds	tens	ones

e. 200 – 143 = _____

hundreds	tens	ones

2. Herman collected 200 shells on the beach. Of those, he kept 136 shells and left the rest on the beach. How many shells did he leave on the beach?

120   Lesson 28:   Subtract from 200 and from numbers with zeros in the tens place.

© 2018 Great Minds®. eureka-math.org

EUREKA MATH

1. Add like units, and record the totals below.

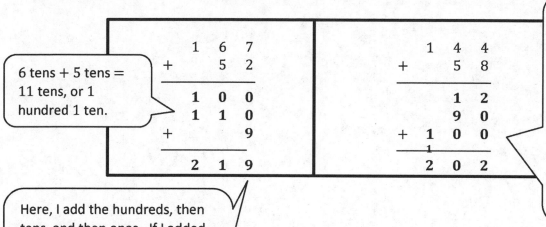

6 tens + 5 tens = 11 tens, or 1 hundred 1 ten.

```
    1  6  7
 +     5  2
 ─────────────
    1  0  0
    1  1  0
 +        9
 ─────────────
    2  1  9
```

```
    1  4  4
 +     5  8
 ─────────────
       1  2
       9  0
 +  1  0  0
    1
 ─────────────
    2  0  2
```

I add all the ones, tens, and hundreds. Look, there are 10 tens! That's the same as 1 hundred 0 tens. I record the hundred on the line.

Here, I add the hundreds, then tens, and then ones. If I added starting with the ones, the totals would still be the same because I am adding the same parts!

2. Dana counted 59 peaches on one tree and 87 peaches on another tree. How many peaches were on both trees? Add like units and record the totals below to solve.

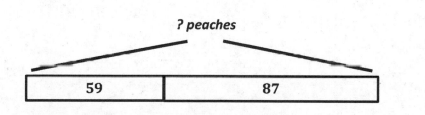

? peaches

| 59 | 87 |

```
       5  9
 +     8  7
 ─────────────
    1  3  0
 +     1  6
 ─────────────
    1  4  6
```

146 *peaches were on both trees.*

EUREKA
MATH

Lesson 29:   Use and explain the totals below method using words, math drawings, and numbers.

© 2018 Great Minds®. eureka-math.org

121

1. Linda and Keith solved 127 + 59.

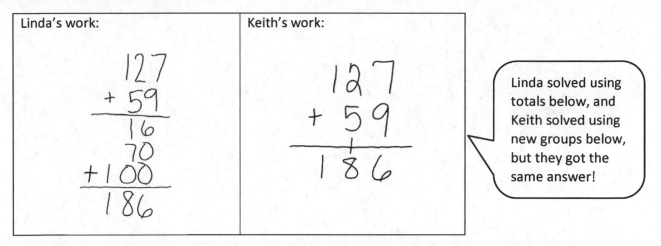

Linda's work:

Keith's work:

Linda solved using totals below, and Keith solved using new groups below, but they got the same answer!

Explain what is different about how Linda and Keith solved the problem.

Linda added the ones, tens, and hundreds by themselves to get the 3 parts: 16, 70, and 100. Then, she added those parts up to get 186. Keith renamed 16 ones as 1 ten 6 ones. Next, he added 2 tens plus 5 tens plus 1 ten, which equals 8 tens. Then, he added 1 hundred. They got the same answer!

2. Here is one way to solve 124 + 69. Solve 124 + 69 another way.

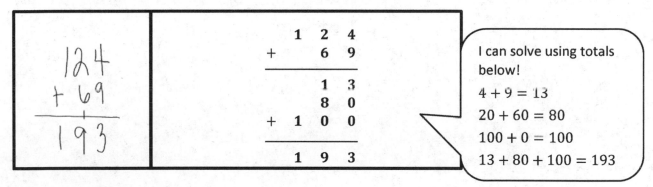

$$
\begin{array}{r}
1\ \ 2\ \ 4 \\
+\ \ \ \ 6\ \ 9 \\
\hline
1\ \ 3 \\
8\ \ 0 \\
+\ 1\ \ 0\ \ 0 \\
\hline
1\ \ 9\ \ 3
\end{array}
$$

I can solve using totals below!
$4 + 9 = 13$
$20 + 60 = 80$
$100 + 0 = 100$
$13 + 80 + 100 = 193$

Explain how the two ways to solve 124 + 69 are similar.

In the first problem, when you rename 13 ones, you can see that 1 hundred 8 tens 13 ones becomes 1 hundred 9 tens 3 ones. When I solve the problem another way, it is just like showing the 3 parts before renaming. 1 hundred 8 tens 13 ones $= 100 + 80 + 13$. I can add the parts in any order and get the same total!

EUREKA MATH

Lesson 30:   Compare totals below to new groups below as written methods.

125

© 2018 Great Minds®. eureka-math.org

Name _____ Date _____

1. Kari and Marty solved 136 + 56.

Kari's work:	Marty's work:
$$\begin{array}{r} 136 \\ + 56 \\ \hline 192 \end{array}$$	$$\begin{array}{r} 136 \\ + 56 \\ \hline 12 \\ 80 \\ +100 \\ \hline 192 \end{array}$$

Explain what is different about how Kari and Marty solved the problem.

_____

_____

_____

2. Here is one way to solve 145 + 67. For (a), solve 145 + 67 another way.

	a.
$\begin{array}{r} 145 \\ +\ 67 \\ \underline{11} \\ 212 \end{array}$	

b. Explain how the two ways to solve 145 + 67 are similar.

_____

_____

_____

3. Show another way to solve 142 + 39.

$\begin{array}{r} 142 \\ +\ 39 \\ \underline{} \\ 11 \\ 70 \\ \underline{100} \\ 181 \end{array}$	

Compare totals below to new groups below as written methods.

© 2018 Great Minds®. eureka-math.org

EUREKA MATH

Solve the following word problems by drawing a tape diagram. Then, use any strategy that you've learned to solve.

Sandra has 46 fewer coins than Martha. Sandra has 57 coins.

a. How many coins does Martha have?

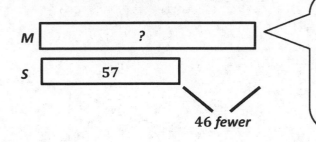

| M | ? |
| S | 57 |

46 fewer

> I use the RDW process to solve. A tape diagram helps me see the parts I know. I know that Sandra has 46 fewer coins than Martha, so that means Martha has more coins, 46 more. I add to find the number of coins Martha has.

$57 + 46 = ?$

3   43

> I use a number bond and the make ten strategy to solve!

$57 + 3 = 60$
$60 + 43 = 103$

Martha has 103 coins.

b. How many coins do Sandra and Martha have together?

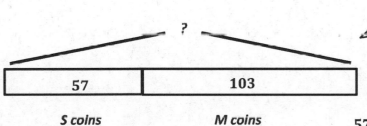

?

| 57 | 103 |

S coins   *M coins*

> From Part (a), I know that Martha has 103 coins. I add them to Sandra's 57 coins to get 160 coins all together.

$57 + 103 = ?$
$57 + 100 + 3 = ?$
$60 + 100 = 160$

Sandra and Martha have 160 coins together.

> I break 103 into $100 + 3$ and then add the 3 to 57 to make the friendly number 60. Now, this problem is easy: $60 + 100 = 160$.

Name _____   Date _____

1. Melissa had 56 pens and 37 more pencils than pens.

   a. How many pencils did Melissa have?

   b. How many pens and pencils did Melissa have?

2. Antonio gave 27 tomatoes to his neighbor and 15 to his brother. He had 72 tomatoes before giving some away. How many tomatoes does Antonio have left?

3. The bakery made 92 muffins. Seventeen were blueberry, 23 were cranberry, and the rest were chocolate chip. How many chocolate chip muffins did the bakery make?

4. After spending $43 on groceries and $19 on a book, Mrs. Groom had $16 left. How much money did Mrs. Groom have to begin with?

EUREKA
MATH

Grade 2
Module 5

1. Complete each *more* or *less* statement.

   a. 10 less than 175 is **165**.

   b. 100 more than 308 is **408**.

   c. **788** is 100 less than 888.

   d. 607 is **10 more** than 597.

   > I can use place value language to explain the change. 10 more and 100 more is the same as adding. 10 less and 100 less is the same as subtracting.

2. Complete each regular number pattern.

   a. 565, 575, **585**, **595**, **605**, 615

   b. 624, **524**, **424**, **324**, **224**, 124, 24

   c. **886**, **876**, **866**, 856, 846, 836

   > I study the numbers and look for the more or less pattern. I know 24 is 100 less than 124, so $24 + 100 = 124$. Then, $124 + 100 = 224$, and so on.

   > I know 846 is 10 less than 856. $856 - 10 = 846$. It's just like taking away a tens disk on the place value chart.

3. Complete each statement.

   a. $609 \xrightarrow{-10} \mathbf{599} \xrightarrow{-100} 499 \xrightarrow{+10} \mathbf{509} \xrightarrow{+10} 519$

   b. $517 \xrightarrow{-10} \mathbf{507} \xrightarrow{-10} \mathbf{497} \xrightarrow{+100} \mathbf{597} \xrightarrow{+10} \mathbf{607} \xrightarrow{+100} \mathbf{707}$

   > I remember the arrow way from Module 4. The arrow way can show a change in the ones, tens, or hundreds place, and it shows whether it's more or less. So, $517 - 10 = 507$. That's a change in the tens place!

4. Solve using the arrow way.

   $\mathbf{220} + 515 = 735$

   $515 \xrightarrow{+100} 615 \xrightarrow{+100} 715 \xrightarrow{+10} 725 \xrightarrow{+10} 735$

   > I start with the part, 515, and add hundreds first until I get to 715. Then, I add tens until I get to 735. $100 + 100 + 10 + 10 = 220$.

Name _____   Date _____

1. Complete each *more* or *less* statement.

a. 10 more than 222 is _____.

b. 100 more than 222 is _____.

c. 10 less than 222 is _____.

d. 100 less than 222 is _____.

e. 515 is 10 more than _____.

f. 299 is 100 less than _____.

g. _____ is 100 less than 345.

h. _____ is 10 more than 397.

i. 898 is _____ than 998.

j. 607 is _____ than 597.

k 10 more than 309 is _____.

l. 309 is _____ than 319.

2. Complete each regular number pattern.

a.   280, 290, _____, _____, _____, 330

b.   530, 520, 510, _____, _____, _____

c.   643, 543 _____, _____, _____, 143

d.   681, 691 _____, _____, _____, 731

e.   427, _____, _____, _____, 387, 377

f.   _____, _____, _____, 788, 778, 768

Lesson 1:   Relate 10 more, 10 less, 100 more, and 100 less to addition and
subtraction of 10 and 100.

137

3. Complete each statement.

   a. $235 \xrightarrow{+10}$ _____ $\xrightarrow{+100}$ _____

   b. $391 \xrightarrow{-100}$ _____ $\xrightarrow{-10}$ _____

   c. $417 \xrightarrow{-10}$ _____ $\xrightarrow{\underline{\quad}}$ _____ $\xrightarrow{-100}$ 297

   d. $311 \xrightarrow{-10}$ _____ $\xrightarrow{-10}$ _____ $\xrightarrow{+100}$ _____ $\xrightarrow{+100}$ _____ $\xrightarrow{+10}$ _____

4. Solve using the arrow way.

   a. 370 + 110 = _____

   b. 290 + _____ = 400

   c. _____ + 710 = 850

Lesson 1:   Relate 10 more, 10 less, 100 more, and 100 less to addition and subtraction of 10 and 100.

© 2018 Great Minds®. eureka-math.org

EUREKA MATH

1. Solve the set of problems using the arrow way.

    $440 + 300 = \underline{\textbf{740}}$

    $440 \xrightarrow{\text{+300}} 740$

    > 300 more than 440 is 740. I just add like units, 4 hundreds plus 3 hundreds is 7 hundreds. The tens and ones stay the same.

    $440 + 360 = \underline{\textbf{800}}$

    $440 \xrightarrow{\text{+300}} 740 \xrightarrow{\text{+60}} 800$

    > To add 360, I add in chunks—hundreds first and then tens. 4 tens + 6 tens = 10 tens, or the next hundred!

    $440 + 380 = \underline{\textbf{820}}$

    $440 \xrightarrow{\text{+300}} 740 \xrightarrow{\text{+60}} 800 \xrightarrow{\text{+20}} 820$

    > The second problem helps me solve this one. 380 is just 20 more than 360. I use the arrow way to add 20. Now, the total is 820.

2. Solve using the arrow way or mental math. Use scrap paper if needed.

    $430 + 290 = \underline{\quad\textbf{720}\quad}$ $660 + 180 = \underline{\quad\textbf{840}\quad}$ $370 + 270 = \underline{\quad\textbf{640}\quad}$

    420   10

    $660 \xrightarrow{\text{+100}} 760 \xrightarrow{\text{+40}} 800 \xrightarrow{\text{+40}} 840$

    > I made a number bond on scrap paper. 290 is close to the next hundred, it just needs 10 more. I broke apart 430 into 420 and 10. I add 10 to 290 and now can solve 420 + 300 in my head.

    > I can solve in my head! 3 hundreds plus 2 hundreds is 5 hundreds. I know 7 tens plus 7 tens is 14 tens, or 140. I can think: 500 + 140 = 640.

    > This is similar to adding 66 and 27, except the units are tens! 6 tens and 7 tens is 13 tens. 60 tens and 20 tens is 80 tens. 13 tens + 80 tens = 93 tens.

    > The first problem can help me solve this one. I notice that 67 tens is 1 more ten than 66 tens. 28 tens is 1 more ten than 27 tens. That means the answer must be 2 more tens than 93 tens!

3. Solve.

    $66 \text{ tens} + 27 \text{ tens} = \underline{\;\textbf{93}\;} \text{ tens}$ $67 \text{ tens} + 28 \text{ tens} = \underline{\;\textbf{95}\;} \text{ tens}$

    What is the value of 85 tens? $\underline{\;\textbf{850}\;}$

3.

b.

c. 3

Name _____ Date _____

1. Solve each set of problems using the arrow way.

a.
260 + 200
260 + 240
260 + 250
b.
320 + 400
320 + 480
320 + 490
c.
550 + 200
550 + 250
550 + 270
d.
230 + 400
230 + 470
230 + 490

2. Solve using the arrow way or mental math. Use scrap paper if needed.

a. 320 + 200 = _____	280 + 320 = _____	290 + 320 = _____
b. 130 + 500 = _____	130 + 560 = _____	130 + 580 = _____
c. 360 + 240 = _____	350 + 270 = _____	380 + 230 = _____
d. 260 + 250 = _____	270 + 280 = _____	280 + 250 = _____
e. 440 + 280 = _____	660 + 160 = _____	770 + 150 = _____

3. Solve.

a.   34 tens + 20 tens = _____ tens   b.   34 tens + 26 tens = _____ tens

c.   34 tens + 27 tens = _____ tens   d.   34 tens + 28 tens = _____ tens

e.   What is the value of 62 tens? _____

1. Solve using the arrow way.

$760 - 400 = \underline{360}$

> I just subtract like units, 7 hundreds minus 4 hundreds is 3 hundreds. The tens and ones stay the same.

$760 \xrightarrow{-400} 360$

$760 - 460 = \underline{300}$

> To subtract 460, I first take away the hundreds and then tens to make it easier!

$760 \xrightarrow{-400} 360 \xrightarrow{-60} 300$

$760 - 480 = \underline{280}$

> The other problems help me solve this one. First, I subtract 400 and then 60 to get to the closest hundred, and now I subtract 20 more. So, I take away 480 in all, one chunk at a time.

$760 \xrightarrow{-400} 360 \xrightarrow{-60} 300 \xrightarrow{-20} 280$

2. Solve using the arrow way or mental math. Use scrap paper if needed.

$640 - 240 = \underline{400}$ $640 - 250 = \underline{390}$ $640 - 290 = \underline{350}$

$640 \xrightarrow{-200} 440 \xrightarrow{-40} 400$ $640 \xrightarrow{-200} 440 \xrightarrow{-40} 400 \xrightarrow{-50} 350$

> I subtract in two steps. First, I take away the hundreds and then the tens. 640 minus 200 is 440. 440 minus 40 is 400.

> I can use the last problem to help me. In my head, I subtract 10 more from 400 since 250 is just 10 more than 240.

> I subtract 290 in chunks: 200, then 40, and then 50. In the last step, I subtract 50 to get to 350.

EUREKA MATH

Lesson 4:   Subtract multiples of 100 and some tens within 1,000.

147

© 2018 Great Minds®. eureka-math.org

> I know that 88 tens minus 20 tens is 68 tens. Then, 68 tens minus 8 tens is 60 tens. Now, I just take away another ten. So, I have 59 tens.

> I could also think of it like this: 88 tens minus 28 tens is 60 tens. Since 29 tens is 1 more than 28 tens, the answer must be 1 less than 60 tens.

> I subtract a total of 28 tens, one chunk at a time. 84 tens minus 20 tens is 64 tens. Now, I take away 4 tens, so I have 60 tens and then 4 more tens, which makes 56 tens.

3.   Solve.

88 tens − 29 tens = __59 *tens*__   84 tens − 28 tens = __56 *tens*__

What is the value of 56 tens? __560__

Lesson 4:   Subtract multiples of 100 and some tens within 1,000.

EUREKA MATH

Name _____   Date _____

1. Solve using the arrow way.

a.
430 – 200
430 – 230
430 – 240

b.
570 – 300
570 – 370
570 – 390

c.
750 – 400
750 – 450
750 – 480

d.
940 – 330
940 – 360
940 – 480

Lesson 4: Subtract multiples of 100 and some tens within 1,000.

149

© 2018 Great Minds®. eureka-math.org

2. Solve using the arrow way or mental math. Use scrap paper if needed.

a.
    330 – 200 = _____   330 – 230 = _____   330 - 260 = _____

b.
    440 – 240 = _____   440 – 260 = _____   440 - 290 = _____

c.
    860 – 560 = _____   860 – 570 = _____   860 – 590 = _____

d.
    970 - 470 = _____   970 – 480 = _____   970 – 490 = _____

3. Solve.

a. 66 tens - 30 tens = _____   b. 66 tens - 36 tens = _____

c. 66 tens - 38 tens = _____   d. 67 tens - 39 tens = _____

e. What is the value of 28 tens? _____

f. What is the value of 36 tens? _____

EUREKA MATH

1. Solve.

> When I have a zero in the ones place, I can think of the number as "some tens"!

43 tens = __430__

24 tens + 19 tens = __43__ tens   25 tens + 29 tens = __54__ tens

> This is similar to 24 + 19 except I am adding tens instead of ones! 19 is just 1 away from 20, so I add 24 tens + 20 tens = 44 tens. Then, I subtract 1 ten and get 43 tens.

> I can use the same idea as the last problem! 25 tens + 30 tens = 55 tens. Since there are only 29 tens, I subtract 1 ten and get 54 tens.

2. Add by drawing a number bond to make a hundred. Write the simplified equation and solve.

   a. $330 + 180$

   310   20

   __310 + 200__ = __510__

   > I can use a number bond to add when one number is close to the next hundred. 180 is close to 200. I need 20 more. I can get it from the 330. I break apart 330 into 310 and 20. Now my problem is $310 + 200$, which is easier to solve. I can just count on 2 hundreds.

   b. $153 + 499$

   152   1

   __152 + 500__ = __652__

   > 499 is only 1 away from 500. I can decompose 153 into 152 and 1. Then, I add the 1 to 499 to get 500. My new addition problem is $152 + 500 = 652$.

   c. $695 + 178$

   5   173

   __700 + 173__ = __873__

   > 695 is closer to the next hundred than 178. I break apart 178 into 5 and 173. I give 5 to 695, so $700 + 173 = 873$.

Name _____ Date _____

1. Solve.

   a. 32 tens = _____   b. 52 tens = _____

   c. 19 tens + 11 tens = _____ tens   d. 19 tens + 13 tens = _____ tens

   e. 28 tens + 23 tens = _____ tens   f. 28 tens + 24 tens = _____ tens

2. Add by drawing a number bond to make a hundred. Write the simplified equation and solve.

   a. 90 + 180
           /\
         10   170

   _____100 + 170_____ = _____

   b. 190 + 460

   _____ = _____

EUREKA MATH

Lesson 5:   Use the associative property to make a hundred in one addend.

© 2018 Great Minds®. eureka-math.org

153

c. 540 + 280

_____ = _____

d. 380 + 430

_____ = _____

e. 99 + 141

_____ = _____

f. 75 + 299

_____ = _____

g. 795 + 156

_____ = _____

Lesson 5:   Use the associative property to make a hundred in one addend.

EUREKA
MATH

1. Draw and label a tape diagram to show how to simplify the problem. Write the new equation, and then subtract.

    a. $570 - 380 =$ ___**590 − 400**___ $=$ ___**190**___

+ 20	570

+ 20	380

    > It's easier to take away hundreds! If I add the same amount, 20, to each number, I have a simpler problem. This is called compensation! Now, I can easily subtract 400 from 590.

    b. $450 - 170 =$ ___**480 − 200**___ $=$ ___**280**___

+ 30	450

+ 30	170

    > I see that 170 is close to 200. I add 30 to each number, so the difference stays the same. My new problem is $480 - 200$.

2. Draw and label a tape diagram to show how to simplify the problem. Write a new equation, and then subtract. Check your work using addition.

    a. $483 - 299 =$ ___**484 − 300**___ $=$ ___**184**___

+ 1	483

+ 1	299

    Check:

    $184 + 300 = 484$

    > I check my work by adding the 2 parts. The sum should be 484.

    > I only need to add 1 to each number to make this problem easier! If I add 1 to both numbers, I can subtract only hundreds, instead of hundreds, tens, and ones!

    > This is much easier than the vertical form because I don't have to rename! I just add 2 to both numbers, and then I can solve in my head!

    b. $776 - 598 =$ ___**778 − 600**___ $=$ ___**178**___

+ 2	776

+ 2	598

    Check:

    $178 + 600 = 778$

EUREKA MATH   **Lesson 6:**   Use the associative property to subtract from three-digit numbers and   155
                         verify solutions with addition.

© 2018 Great Minds®. eureka-math.org

Name _____   Date _____

1. Draw and label a tape diagram to show how to simplify the problem. Write the new equation, and then subtract.

   a. 340 - 190 = <u>350 - 200</u> = _____

+ 10	340

+ 10	190

   b. 420 - 190 = _____ = _____

   c. 500 - 280 = _____ = _____

   d. 650 - 280 = _____ = _____

   e. 740 - 270 = _____ = _____

 EUREKA MATH®   **Lesson 6:**   Use the associative property to subtract from three-digit numbers and   157
verify solutions with addition.

© 2018 Great Minds®. eureka-math.org

2. Draw and label a tape diagram to show how to simplify the problem. Write a new equation, and then subtract. Check your work using addition.

a. 236 - 99 = __237-100__ = _____

+1 | 236 +1 | 99	Check:

b. 372 - 199 = _____ = _____

	Check:

c. 442 - 298 = _____ = _____

	Check:

d. 718-390 = _____ = _____

	Check:

Lesson 6:   Use the associative property to subtract from three-digit numbers and verify solutions with addition.

© 2018 Great Minds®. eureka-math.org

EUREKA MATH

1. Solve each problem with a written strategy such as a tape diagram, a number bond, the arrow way, the vertical form, or chips on a place value chart.

   780 − 390 = **390**   **331** + 600 = 931   **280** = 560 − 280

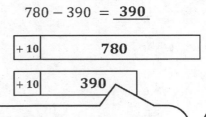

   | + 10 | **780** |
   | + 10 | **390** |

   390 is only 10 away from 400. I draw a tape diagram to show how I add 10 to both numbers so the difference stays the same. My new problem is 790 − 400 = 390.

   +300   +31
   600 → 900 → 931

   I use the arrow way to add in chunks. First, I add 3 hundreds to get to 900. Then, I add 31 more to get to 931. 300 + 31 = 331

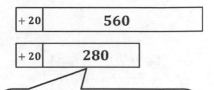

   | + 20 | **560** |
   | + 20 | **280** |

   I use compensation and add 20 to both numbers. So, my easier problem is 580 − 300 = 280. I don't have to unbundle a hundred!

2. Use the arrow way to complete the number sentence.

   820 − 340 = **480**

   $$820 \xrightarrow{-300} 520 \xrightarrow{-20} 500 \xrightarrow{-20} 480$$

   I use the arrow way to subtract hundreds and then tens. When I get to 520, I subtract 20 to get to the hundred and then 20 more to get to 480.

3. Solve 447 + 398 using two different strategies.

   a.
   447 + 398 = 845

   $$447 \xrightarrow{+300} 747 \xrightarrow{+3} 750 \xrightarrow{+50} 800 \xrightarrow{+40} 840 \xrightarrow{+5} 845$$

   b.
   447 + 398 = **845**

   445   2

   445 + 400 = 845

   c. Explain which strategy is easier to use when solving and why.

   It is much easier for me to solve with a number bond because 398 is only 2 away from the next hundred. The arrow way takes a long time, and I have to make sure I don't miss any parts of the number 398. The number bond has a lot fewer steps!

Name _____   Date _____

1. Solve each problem with a written strategy such as a tape diagram, a number bond, the arrow way, the vertical form, or chips on a place value chart.

a.	b.	c.
370 + 300 = _____	_____ = 562 - 200	_____ + 500 = 812
d.	e.	f.
230 - 190 = _____	_____ = 640 - 180	450 - 290 = _____

2. Use the arrow way to complete the number sentences.

a.	b.	c.
420 – 230 = _____	340 – 160 = _____	710 – 350 = _____

EUREKA MATH

Lesson 7:   Share and critique solution strategies for varied addition and subtraction problems within 1,000.

161

© 2018 Great Minds®. eureka-math.org

3. Solve 667 + 295 using two different strategies.

a.	b.

    c. Explain which strategy is easier to use when solving and why.

_____

_____

_____

4. Circle one of the strategies below, and use the circled strategy to solve 199 + 478.

a.	b. Solve:
arrow way / number bond	

    c. Explain why you chose that strategy.

_____

_____

_____

Lesson 7:   Share and critique solution strategies for varied addition and subtraction problems within 1,000.

© 2018 Great Minds®. eureka-math.org

EUREKA MATH

1. Solve the following problems using your place value chart, place value disks, and vertical form. Bundle a ten or hundred when necessary.

516 + 224

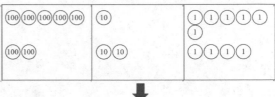

```
    5  1  6
 +  2  2  4
```

> I write the problem in vertical form and model both addends with my place value disks.

```
    5  1  6
 +  2  2  4
       1
          0
```

> 6 ones plus 4 ones is 10 ones, or 1 ten 0 ones. I record this in vertical form on the line below the tens place by first showing the new unit of ten using new groups below. Then, I write 0 below the ones place.

```
    5  1  6
 +  2  2  4
       1
    7  4  0
```

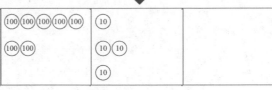

> Next, I add the tens and then the hundreds. 1 ten plus 2 tens plus 1 more ten is 4 tens. 5 hundreds plus 2 hundreds is 7 hundreds.
>
> $516 + 224 = 740$

2. Solve.

a. $600 + 180 = \underline{780}$

> Easy! $600 + 100 = 700$. Then I add on 80, so 780.

b. $620 + 180 = \underline{800}$
   /\
   600 20

> I can break 620 into 600 and 20 to make an easier problem to solve. When I add the 20 to 180 I get 200, and $600 + 200 = 800$, so $620 + 180 = 800$.

c. $680 + 220 = \underline{900}$
   \/\
   800 100

> $600 + 200 = 800$
> $80 + 20 = 100$
> $800 + 100 = 900$

d. $680 + 230 = \underline{910}$

> Part (c) helps me solve this one. The first addend, 680, is the same. 230 is just 10 more than 220. That means the answer must be 10 more than 900, or 910.

Name _____   Date _____

1. Solve the following problems using your place value chart, place value disks, and
    vertical form. Bundle a ten or hundred, when necessary.

a. 505 + 75	b. 606 + 84
c. 293 + 114	d. 314 + 495
e. 364 + 326	f. 346 + 234
g. 384 + 225	h. 609 + 351

2. Solve.

a. 200 + 400 = _____

b. 220 + 400 = _____

c. 220 + 440 = _____

d. 220 + 480 = _____

e. 225 + 485 = _____

f. 500 + 60 = _____

g. 500 + 160 = _____

h. 540 + 160 = _____

i. 560 + 240 = _____

j. 560 + 250 = _____

Lesson 8:   Relate manipulative representations to the addition algorithm.

EUREKA MATH

1. Solve the following problems using your place value chart, place value disks, and vertical form. Bundle a ten or hundred when necessary.

$$346 + 278$$

I show each step with the place value disks in the vertical form. When I make a new unit, I show it with new groups below.

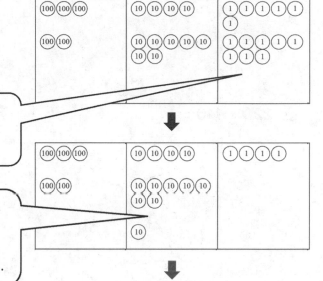

$$
\begin{array}{r}
3\ 4\ 6 \\
+\ 2\ 7\ 8 \\
\hline
{\scriptstyle 1} \\
4
\end{array}
$$

When I add the ones, I have 14 ones, or 1 ten 4 ones. I change 10 ones for 1 ten.

$$
\begin{array}{r}
3\ 4\ 6 \\
+\ 2\ 7\ 8 \\
\hline
{\scriptstyle 1}\ {\scriptstyle 1} \\
2\ 4
\end{array}
$$

Next, I add 4 tens plus 7 tens plus 1 more ten. That's 12 tens, or 1 hundred 2 tens. I change 10 tens for 1 hundred.

$$
\begin{array}{r}
3\ 4\ 6 \\
+\ 2\ 7\ 8 \\
\hline
{\scriptstyle 1}\ {\scriptstyle 1} \\
6\ 2\ 4
\end{array}
$$

Now I have 6 hundreds 2 tens 4 ones. $346 + 278 = 624$

2. Solve.

a. $478 + 303 =$ __781__

    $\overset{\wedge}{\underset{2\ \ \ 301}{}}$

478 is close to 480; it only needs 2 more. I can take 2 from 303 by breaking 303 into 2 and 301 to make an easier problem. $480 + 301 = 781$, so $478 + 303 = 781$.

b. $478 + 323 =$ __801__

Part (a) helps me solve this problem. 323 is just 20 more than 303, so the answer must be 20 more than 781. I count on 2 tens from 781. $781, 791, 801$.

Name _____   Date _____

1. Solve the following problems using a place value chart, place value disks, and vertical form. Bundle a ten or hundred, when necessary.

a. 205 + 345	b. 365 + 406
c. 446 + 334	d. 466 + 226
e. 537 + 243	f. 358 + 443
g. 753 + 157	h. 663 + 258

2. Solve.

   a. 180 + 420 = _____

   b. 190 + 430 = _____

   c. 364 + 236 = _____

   d. 275 + 435 = _____

   e. 404 + 206 = _____

   f. 440 + 260 = _____

   g. 444 + 266 = _____

EUREKA MATH

Solve using vertical form, and draw chips on the place value chart. Bundle as needed.

$306 + 596 = \underline{\mathbf{902}}$

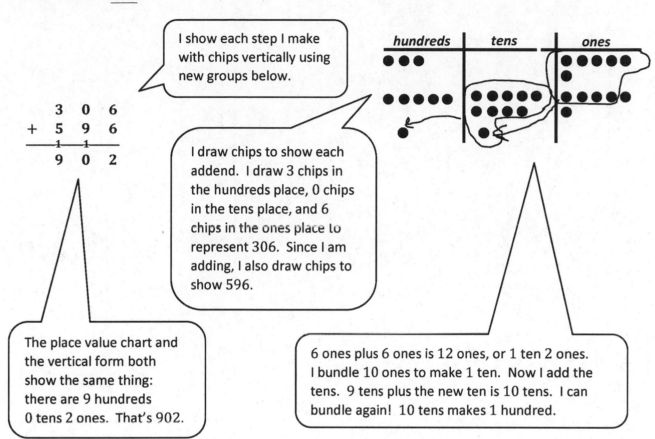

I show each step I make with chips vertically using new groups below.

$$
\begin{array}{r}
3\ \ 0\ \ 6 \\
+\ 5\ \ 9\ \ 6 \\
\hline
{}_1\ \ {}_1\quad \\
9\ \ 0\ \ 2 \\
\end{array}
$$

I draw chips to show each addend. I draw 3 chips in the hundreds place, 0 chips in the tens place, and 6 chips in the ones place to represent 306. Since I am adding, I also draw chips to show 596.

The place value chart and the vertical form both show the same thing: there are 9 hundreds 0 tens 2 ones. That's 902.

6 ones plus 6 ones is 12 ones, or 1 ten 2 ones. I bundle 10 ones to make 1 ten. Now I add the tens. 9 tens plus the new ten is 10 tens. I can bundle again! 10 tens makes 1 hundred.

EUREKA
MATH

Lesson 10:   Use math drawings to represent additions with up to two   171
              compositions and relate drawings to the addition algorithm.

© 2018 Great Minds®. eureka-math.org

Solve using vertical form, and draw chips on the place value chart. Bundle as needed.

$276 + 324 =$ ___**600**___

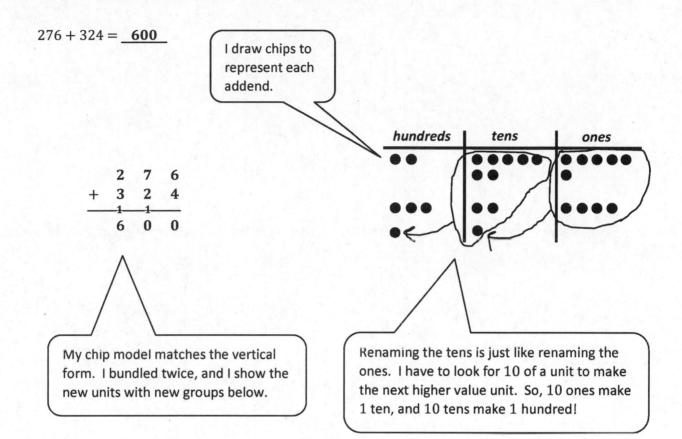

I draw chips to represent each addend.

$$
\begin{array}{r}
2\ \ 7\ \ 6 \\
+\ 3\ \ 2\ \ 4 \\
{}_{1}\ \ {}_{1}\ \ \ \\
\hline
6\ \ 0\ \ 0
\end{array}
$$

My chip model matches the vertical form. I bundled twice, and I show the new units with new groups below.

Renaming the tens is just like renaming the ones. I have to look for 10 of a unit to make the next higher value unit. So, 10 ones make 1 ten, and 10 tens make 1 hundred!

Lesson 11: Use math drawings to represent additions with up to two compositions and relate drawings to the addition algorithm.

175

1. Solve 246 + 490 using two different strategies.

a. **246 + 490 = 736**

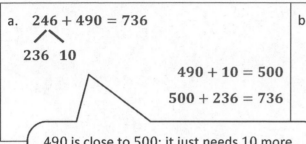

236 10

490 + 10 = 500

500 + 236 = 736

490 is close to 500; it just needs 10 more, so I make the next hundred by breaking 246 into 236 and 10. This is the easiest strategy because it's easy to add 5 hundreds to 236.

b.

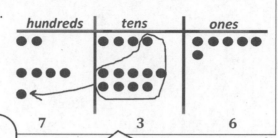

hundreds	tens	ones
7	3	6

I could also draw a chip model, but that would take longer, so it's not as efficient as using a number bond.

2. Choose the best strategy and solve. Explain why you chose that strategy.

a. 499 + 367 = **866**

1   366

The best strategy is to make the next hundred to make an easier problem to solve. 499 needs just 1 more to be 500. Then, it's easy to add what's left, 366. 500 + 366 = 866, so 499 + 367 = 866. That's why it's important to always look for relationships between the numbers.

b. 534 + 110 = **644**

I can solve this one mentally by adding like units. 500 + 100 = 600, and 34 + 10 = 44, so 600 + 44 = 644.

c. 695 + 248 = **943**

5   243

At first, I thought I needed to use the chip model and vertical form because I can see I need to rename twice. But then I looked more carefully! I see that I can make the next hundred, so I break apart 248. 695 + 5 = 700, and 700 + 243 = 943, so 695 + 248 = 943.

EUREKA MATH

Lesson 12: Choose and explain solution strategies and record with a written addition method.

179

Name _____ Date _____

1. Solve 435 + 290 using two different strategies.

a.	b.

c. Explain which strategy would be easier and why.

_____

_____

_____

_____

EUREKA MATH

Lesson 12: Choose and explain solution strategies and record with a written addition method.

© 2018 Great Minds®. eureka-math.org

181

2. Choose the best strategy and solve. Explain why you chose that strategy.

a. 299 + 458	Explanation:
b. 733 + 210	Explanation:
c. 295 + 466	Explanation:

Lesson 12:   Choose and explain solution strategies and record with a written
                addition method.

© 2018 Great Minds®. eureka-math.org

EUREKA
MATH

I can use $180 - 30$ to help me solve $180 - 29$. Since the difference in the first problem is 150, the difference in the second problem must be 1 more than 150 because I am subtracting 1 less.

1. Solve using mental math.

$8 - 3 = \underline{\ \ 5\ \ }$ $80 - 30 = \underline{\ \ 50\ \ }$ $180 - 30 = \underline{\ \ 150\ \ }$ $180 - 29 = \underline{\ \ 151\ \ }$

2. Solve using mental math or vertical form with place value disks. Check your work using addition.

   a. $223 - 121 = \underline{\ \ 102\ \ }$

I can use mental math to solve because there's no renaming. I just subtract like units. $200 - 100 = 100$, $20 - 20 = 0$, and $3 - 1 = 2$. $100 + 2 = 102$, so $223 - 121 = 102$. I can check my work by adding: $102 + 121 = 223$.

   b. $378 - 119 = \underline{\ \ 259\ \ }$

I can solve this one mentally, too, using compensation. If I add 1 to each number, I make a problem that's easier to solve, $379 - 120$. There's no renaming, so I just subtract like units. The answer is 259.

+1	378

+1	119

I know that part plus part equals whole, so if I'm right, $259 + 119$ must equal 378. When I check my work, I see that I'm right!

$$\begin{array}{r} 2\ \ 5\ \ 9 \\ +\ 1\ \ 1\ \ 9 \\ \hline {\scriptstyle 1} \\ 3\ \ 7\ \ 8 \end{array}$$

3. Complete the number sentence modeled by place value disks.

The model shows the whole, 342. 2 hundreds 2 tens 5 ones are crossed off. That's 225. That means the number sentence is $342 - 225 = 117$. I can check to see if I'm right by adding 117 and 225.

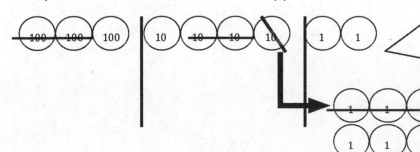

$$\underline{\ \ 342\ \ } - \underline{\ \ 225\ \ } = 117$$

$$\begin{array}{r} 1\ \ 1\ \ 7 \\ +\ 2\ \ 2\ \ 5 \\ \hline {\scriptstyle 1} \\ 3\ \ 4\ \ 2 \end{array}$$

EUREKA MATH

Lesson 13:   Relate manipulative representations to the subtraction algorithm, and use addition to explain why the subtraction method works.

183

© 2018 Great Minds®. eureka-math.org

Name _____   Date _____

1. Solve using mental math.

    a. 9 – 5 = _____   90 – 50 = _____   190 – 50 = _____   190 – 49 = _____

    b. 7 – 4 = _____   70 – 40 = _____   370 – 40 = _____   370 – 39 = _____

2. Solve using mental math or vertical form with place value disks. Check your work using addition.

    a. 53 – 31 = ___122___   b. 153 – 38 = _____

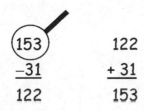

         153   122
        – 31   + 31
         122   153

    c. 362 – 49 = _____   d. 485 – 177 = _____

EUREKA
MATH

Lesson 13:   Relate manipulative representations to the subtraction algorithm, and
             use addition to explain why the subtraction method works.

© 2018 Great Minds®. eureka-math.org

185

e. 753 – 290 = _____

f. 567 – 290 = _____

g. 873 – 428 = _____

h. 817 – 565 = _____

i. 973 – 681 = _____

j. 748 – 239 = _____

3. Complete the number sentence modeled by place value disks.

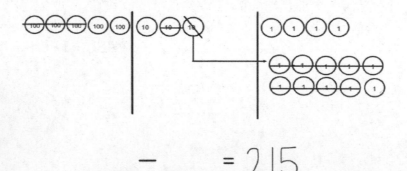

_____ – _____ = 215

Lesson 13:   Relate manipulative representations to the subtraction algorithm, and use addition to explain why the subtraction method works.

© 2018 Great Minds®. eureka-math.org

EUREKA MATH

1. Solve by drawing place value disks on a chart. Then, use addition to check your work.

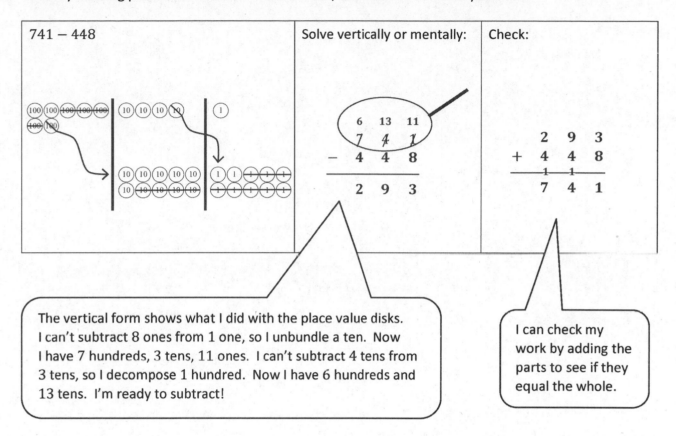

| 741 − 448 | Solve vertically or mentally: | Check: |

The vertical form shows what I did with the place value disks. I can't subtract 8 ones from 1 one, so I unbundle a ten. Now I have 7 hundreds, 3 tens, 11 ones. I can't subtract 4 tens from 3 tens, so I decompose 1 hundred. Now I have 6 hundreds and 13 tens. I'm ready to subtract!

I can check my work by adding the parts to see if they equal the whole.

2. If 584 − 147 = 437, then 437 + 147 = 584. Explain why this statement is true using numbers, pictures, or words.

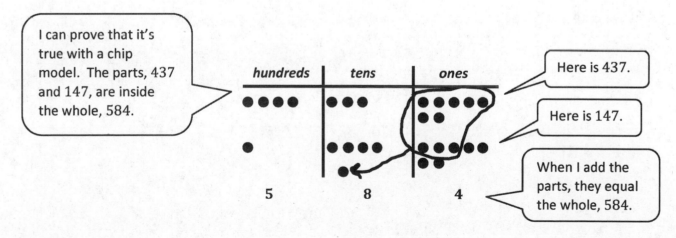

I can prove that it's true with a chip model. The parts, 437 and 147, are inside the whole, 584.

Here is 437.

Here is 147.

When I add the parts, they equal the whole, 584.

EUREKA MATH

Lesson 14:   Use math drawings to represent subtraction with up to two
              decompositions, relate drawings to the algorithm, and use addition to
              explain why the subtraction method works.
© 2018 Great Minds®. eureka-math.org

187

Name _____   Date _____

1. Solve by drawing place value disks on a chart. Then, use addition to check your work.

	Solve vertically or mentally:	Check:
a. 373 – 180		
b. 463 – 357		
c. 723 – 584		

EUREKA MATH

Lesson 14:   Use math drawings to represent subtraction with up to two decompositions, relate drawings to the algorithm, and use addition to explain why the subtraction method works.

© 2018 Great Minds®. eureka-math.org

189

d. 861 – 673	Solve vertically or mentally:	Check:
e. 898 – 889	Solve vertically or mentally:	Check:

2. If 544 + 366 = 910, then 910 - 544 = 366. Explain why this statement is true using numbers, pictures, or words.

Lesson 14:   Use math drawings to represent subtraction with up to two decompositions, relate drawings to the algorithm, and use addition to explain why the subtraction method works.

EUREKA
MATH

1. Solve by drawing chips on the place value chart. Then, use addition to check your work.

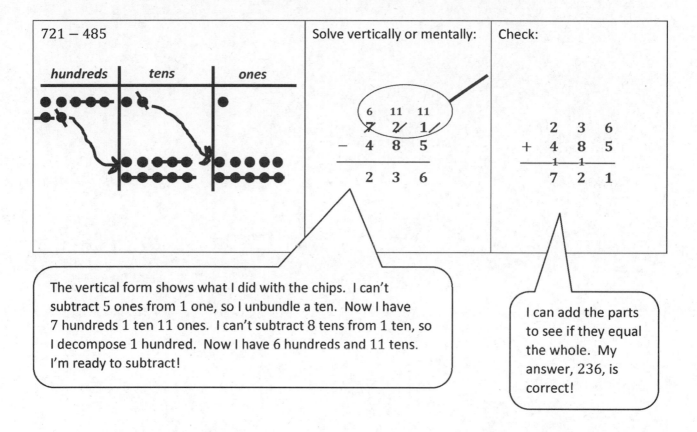

| 721 − 485 | Solve vertically or mentally: | Check: |

The vertical form shows what I did with the chips. I can't subtract 5 ones from 1 one, so I unbundle a ten. Now I have 7 hundreds 1 ten 11 ones. I can't subtract 8 tens from 1 ten, so I decompose 1 hundred. Now I have 6 hundreds and 11 tens. I'm ready to subtract!

I can add the parts to see if they equal the whole. My answer, 236, is correct!

2. Complete the *if...then* statement. Draw a number bond to represent the related facts.

If 631 − __**358**__ = 273, then __**358**__ + 273 = 631.

The number bond shows the part–whole relationship.

I know that whole − part = part. 631 is the whole because it's the largest number. 273 is the part I know, so I can subtract to find the other part: 631 − 273 = 358. That also means that 358 + 273 = 631 because part + part = whole.

$$\begin{array}{ccc} 5 & 12 & 11 \\ 6 & 3 & 1 \\ -\ 2 & 7 & 3 \\ \hline 3 & 5 & 8 \end{array}$$

631
273   358

1. Solve vertically or using mental math. Draw chips on the place value chart and unbundle if needed.

a. $408 - 261 = \underline{\textbf{147}}$

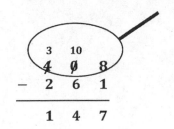

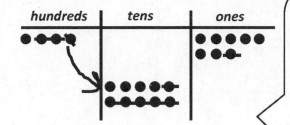

hundreds	tens	ones

> The vertical form shows what I did with the chips. I have enough ones to subtract in the ones place, but I need to unbundle 1 hundred to have enough tens in the tens place. Now I'm ready to subtract!

b. $700 - 568$ $\underline{\textbf{132}}$

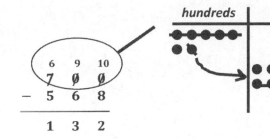

hundreds	tens	ones

> I see that both the tens and the ones are going to need more. I can unbundle a hundred in one step. 1 hundred is equal to 9 tens 10 ones. Now I have 6 hundreds 9 tens 10 ones. I show this with my chips and in the vertical form. Now, I am ready to subtract.

2. Emily said that $400 - 247$ is the same as $399 - 246$. Write an explanation using pictures, numbers, or words to prove Emily is correct.

> I can explain two different ways!

> I can use compensation. I notice that 400 is just 1 more than 399, and 247 is just 1 more than 246. So, the difference for each problem must be the same!

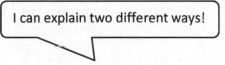

+1	399

+1	246

> I can use the arrow way to show that the difference is the same. $400 - 247 = 153$, and $399 - 246 = 153$.

$$400 \xrightarrow{-200} 200 \xrightarrow{-40} 160 \xrightarrow{-7} 153$$

$$399 \xrightarrow{-200} 199 \xrightarrow{-40} 159 \xrightarrow{-6} 153$$

Lesson 16: Subtract from multiples of 100 and from numbers with zero in the tens place.

195

EUREKA MATH®

Name _____   Date _____

1. Solve vertically or using mental math. Draw chips on the place value chart and unbundle, if needed.

a. 206 – 89 = _____

hundreds	tens	ones

b. 509 – 371 = _____

hundreds	tens	ones

c. 607 – 288 = _____

hundreds	tens	ones

EUREKA MATH

Lesson 16:   Subtract from multiples of 100 and from numbers with zero in the tens place.

© 2018 Great Minds®. eureka-math.org

197

d. 800 – 608 = _____

hundreds	tens	ones

e. 900 – 572 = _____

hundreds	tens	ones

2. Andy said that 599 – 456 is the same as 600 – 457. Write an explanation using pictures, numbers, or words to prove Andy is correct.

Lesson 16:   Subtract from multiples of 100 and from numbers with zero in the tens place.

EUREKA MATH

© 2018 Great Minds®. eureka-math.org

Solve vertically or using mental math. Draw chips on the place value chart and unbundle if needed.

a.   500 – 231   __269__

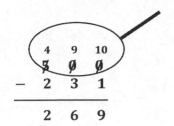

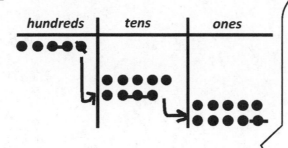

I see that both the tens and the ones are going to need more. I can unbundle a hundred in one step. 1 hundred is equal to 9 tens 10 ones. Now I have 4 hundreds 9 tens 10 ones. I show this with my chips and in the vertical form. I am ready to subtract.

b.   902 - 306 = __596__

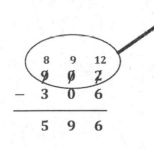

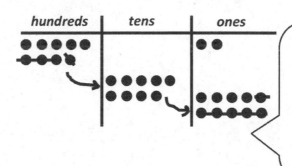

I change 1 hundred for 9 tens 10 ones. Now I have 8 hundreds 9 tens 12 ones. I show my work with the chips and in the vertical form. I am ready to subtract.

I can check my work using addition.

596 + 306 = 902

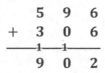

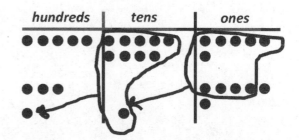

EUREKA MATH

Lesson 17:   Subtract from multiples of 100 and from numbers with zero in the tens place.

199

© 2018 Great Minds®. eureka-math.org

1. Use the arrow way and counting on to solve.

$300 - 164 = 136$ $164 \xrightarrow{+6} 170 \xrightarrow{+30} 200 \xrightarrow{+100} 300$

The arrow way is efficient. I add ones, tens, and hundreds to get to benchmark, or friendly, numbers. That makes counting on easy!

I added $6 + 30 + 100$. That equals 136. So $300 - 164 = 136$.

2. Choose a strategy to solve, and explain why you chose that strategy.

$500 - 280 = 220$

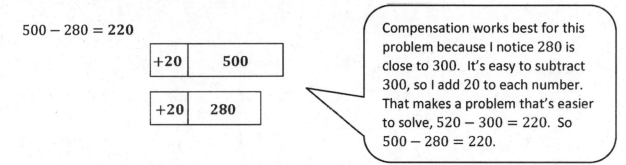

Compensation works best for this problem because I notice 280 is close to 300. It's easy to subtract 300, so I add 20 to each number. That makes a problem that's easier to solve, $520 - 300 = 220$. So $500 - 280 = 220$.

3. Explain why $400 - 173$ is the same as $399 - 172$.

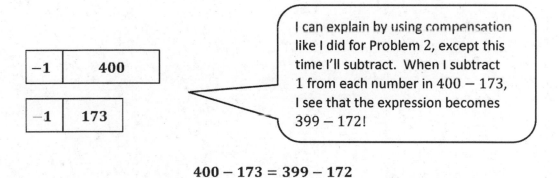

I can explain by using compensation like I did for Problem 2, except this time I'll subtract. When I subtract 1 from each number in $400 - 173$, I see that the expression becomes $399 - 172$!

$$400 - 173 = 399 - 172$$

EUREKA
MATH

Lesson 18: Apply and explain alternate methods for subtracting from multiples of 100 and from numbers with zero in the tens place.

© 2018 Great Minds®. eureka-math.org

203

Name _____   Date _____

1. Use the arrow way and counting on to solve.

a. 700 – 462	b. 900 – 232

2. Solve vertically, and draw a place value chart and chips. Rename in one step.

a. 907 – 467	b. 803 – 667

3. Choose a strategy to solve, and explain why you chose that strategy.

a. 700 – 390	Explanation:

Lesson 18:   Apply and explain alternate methods for subtracting from multiples of
100 and from numbers with zero in the tens place.

© 2018 Great Minds®. eureka-math.org

205

b. 919 − 657	Explanation:

4. Explain why 300 − 186 is the same as 299 − 185.

Explanation:

5. Solve 500 - 278 usina the simplifying strategy from Problem 4.

Solution:

Lesson 18:   Apply and explain alternate methods for subtracting from multiples of
                  100 and from numbers with zero in the tens place.

© 2018 Great Minds®. eureka-math.org

EUREKA
MATH

Solve and explain why you chose that strategy.

a. $580 + 230 = \underline{\textbf{810}}$
              $\wedge$
           20 210

> I notice I can make the next hundred because 580 is close to 600. I break apart 230 into 20 and 210. 600 more than 210 is easy, 810.

b. $310 + \underline{\textbf{333}} = 643$

   $643 - 310 = 333$

> To find a missing addend, I can subtract. If I subtract one part from the whole, the answer is the missing part. I rewrite the problem as $643 - 310$. There's no renaming so I just subtract like units, hundreds from hundreds, tens from tens, and ones from ones.

c. $900 - 327 = \underline{\textbf{573}}$

   $327 \xrightarrow{+3} 330 \xrightarrow{+70} 400 \xrightarrow{+500} 900$

> The arrow way is easy because I just need to reach a benchmark number, and then I can skip-count quickly. $327 + 3$ gets me to 330. 330 needs 70 to get to 400. Now I just add 500 to reach 900. Altogether, I added 573, so $900 - 327 = 573$.

d. $802 - 698$ $\underline{\textbf{104}}$

+2	802

+2	698

> I can use compensation. I notice that 698 is very close to 700, which is an easy number to subtract. I add 2 to 698 to get 700. What I do to one number I must do to the other number, so I add 2 to 802. Now I have an easier problem to solve, $804 - 700$. Easy! The answer is 104.

Name _____ Date _____

1. Solve and explain why you chose that strategy.

a. 340 + 250 = _____	Explanation: _____ _____ _____ _____
b. 490 + 350 = _____	Explanation: _____ _____ _____ _____ _____
c. 519 + 342 = _____	Explanation: _____ _____ _____ _____

Lesson 19:   Choose and explain solution strategies and record with a written addition
             or subtraction method.

© 2018 Great Minds®. eureka-math.org

209

d. 610 + _____ = 784	Explanation: _____ _____ _____ _____
e. 700 – 456 = _____	Explanation: _____ _____ _____ _____
f. 904 – 395 = _____	Explanation: _____ _____ _____ _____

Lesson 19:   Choose and explain solution strategies and record with a written addition
              or subtraction method.

© 2018 Great Minds®. eureka-math.org

EUREKA
MATH

1. Solve each problem using two different strategies.

295 + __239__ = 534

a. First Strategy	b. Second Strategy

a. First Strategy

$$295 \xrightarrow{+5} 300 \xrightarrow{+200} 500 \xrightarrow{+34} 534$$

> I can solve by counting on. I use the arrow way to show what I add to 295 to reach 534.
>
> $200 + 34 + 5 = 239$

b. Second Strategy

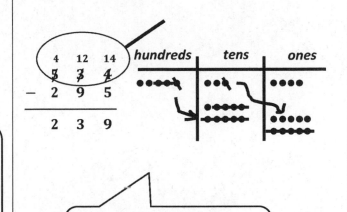

> I can also solve by using a chip model, and I show my work in vertical form.

2. Circle a strategy to solve and explain why you chose that strategy.

843 – 698 = __145__   *Number bond or arrow way*

$$698 \xrightarrow{+2} 700 \xrightarrow{+100} 800 \xrightarrow{+43} 843$$

> I chose the arrow way because I see that 698 is close to 700. I just add 2. From there, I can add 100 to reach 800. Then I just add 43 more to reach 843. $100 + 43 + 2 = 145$

EUREKA MATH

Lesson 20: Choose and explain solution strategies and record with a written addition or subtraction method.

211

© 2018 Great Minds®. eureka-math.org

Name _____   Date _____

Solve each problem using two different strategies.

1. 456 + 244 = _____

a. First Strategy	b. Second Strategy

2. 698 + _____ = 945

a. First Strategy	b. Second Strategy

Circle a strategy to solve, and explain why you chose that strategy.

3. 257 + 160 = _____

        a. *Arrow way or vertical form*

b. Solve:	c. Explanation:
	_____
	_____
	_____
	_____
	_____

4. 754 – 597 = _____

        a. *Number bond or arrow way*

b. Solve:	c. Explanation:
	_____
	_____
	_____
	_____
	_____

Lesson 20:   Choose and explain solution strategies and record with a written addition
              or subtraction method.

EUREKA
MATH

Credits

Great Minds® has made every effort to obtain permission for the reprinting of all copyrighted material. If any owner of copyrighted material is not acknowledged herein, please contact Great Minds for proper acknowledgment in all future editions and reprints of this module.